Changing Business from the Inside Out:
A Treehugger's Guide to Working in Corporations

"Tim Mohin's career has shown how business activity can create great outcomes for the world, as well as for companies. Now he has provided an essential roadmap for anyone who wants to shape a career that shapes the world. *Changing Business from the Inside Out* shows how to make real change happen – and should be essential reading for anyone who wants to build a meaningful career."

Aron Cramer, CEO, Business for Social Responsibility

"A compelling behind-the-scenes guide to being an effective internal change agent, this book is a must-read for any recent MBA graduate or business professional entering the field of corporate responsibility. Companies are critical market actors in our overdue transition to a sustainable global economy, and *Changing Business from the Inside Out* demonstrates how you can leverage your passion and expertise to implement a strong CR program and sustainable business strategy."

Mindy Lubber, President and CEO, Ceres

"This is the ultimate insider's guide, from someone who has been at the front lines of corporate change-making at some of the world's biggest companies. Whether you are starting your career or seeking to infuse your current one with meaning and purpose, whether you're a business leader or simply hope to become one, this is the roadmap you'll need to succeed."

Joel Makower, publisher of GreenBiz.com

"At a time when more and more people are entering the workforce with the intention of making a contribution to people and the planet, Tim Mohin has written a timely and important book. *Changing Business from the Inside Out* provides practical, hands-on advice gleaned from nearly 30 years of experience. As a Net Impact board member and an experienced business leader, Tim is able to seamlessly weave the critical needs of the business with the desires of today's young people into a valuable guide for a career with impact."

Liz Maw, Executive Director Net Impact

"So many business books are written from the outside looking in. As an executive at Intel, Apple, and AMD, Tim Mohin has seen corporate America from the inside – and he has done a masterful job of turning his experience into a readable, insightful, and valuable guide for anyone who wants to use the power of business to change the world for the better. Bravo!"

Marc Gunther, contributing editor at *Fortune* and senior writer at GreenBiz.com

Management guru Peter Drucker challenged those in business to look beyond conventional measures of success and to ask themselves, 'What do you want to be remembered for?' For those in business or contemplating a career in the private sector, Tim Mohin offers a user's guide and a constructive toolkit for successfully pursuing a path to doing good while doing well. Written by a pioneering practitioner who shares lessons from the trenches, Tim not only shares his valuable experience as an environmental champion but also supplies a detailed and disciplined approach to achieving results from within the corporation. He proves that business, when done right, can indeed be an instrument for good, and that a new generation of passionate and competent managers can change the corporate DNA in a way that achieves value for shareholders and society alike. This is a book not for dreamers but for doers – doers who want to produce quantifiable and meaningful results. Applying Tim's lessons and techniques offers those who accept Drucker's challenge to leave a legacy of doing good work that offers both professional fulfillment and lasting value for society.

Ira A. Jackson, Distinguished Scholar, MIT Legatum Center for Development and Entrepreneurship; former Dean, Peter Drucker and Masatoshi Ito School of Management, Claremont Colleges; co-author (with Jane Nelson) of *Profits with Principles: Seven Strategies for Delivering Value with Values*

The line between a 'treehugger' and a hard-nosed, profit-seeking businessperson is getting very blurry. Tim Mohin walks right on that line. The transformation of business to sustainable operations has begun, and all companies need to learn from those few and proud executives who have been at the front lines for years. When I set out to interview experts for my book *Green to Gold*, Tim was one of my first stops. His new book will show you why.

Andrew Winston, sustainability strategist; co-author of *Green to Gold*

Mohin draws upon his 20 years of experience to deliver a book that is equally inspiring and practical. *Changing Business from the Inside Out: A Treehugger's Guide to Working in Corporations* comes at a time when expectations for for-profit companies to do the right thing have never been higher and is a must-read for anyone working a corporate job, as change is all of our responsibility.

Darrell Hammond, Founder and CEO, KaBOOM!
NYT bestselling author of *KaBOOM! How One Man Built a Movement to Save Play*

"There are many ways to be a force for corporate responsibility and sustainability, but none more powerful than when from within companies that matter. Tim Mohin's bold new book takes us with him into the belly of the beast and as he aligns progressive values with business interests in ways that improve the world. Mohin's message is that it can be done – and he shows us how."

Bennett Freeman, Senior Vice President for Sustainability Research and Policy, Calvert Investments

"Tim's work focuses squarely on the much-needed sustainability unlock: showing each and every one of us how we can embed it in our everyday lives, leverage the power of ourselves, and the organizations for which we work, to change the world. Perfect and powerful."

Dr. Kellie A. McElhaney, Alexander Faculty Fellow in Corporate Responsibility, Center for Responsible Business, Haas School of Business at University of California, Berkeley

"More than a roadmap for would-be sustainability professionals, Mohin shows any business leader – or anyone looking to influence companies – that to make a lasting impact CSR efforts must align with a firm's core values, tactics must be vetted by key internal and external stakeholders, and benefits must ring true with customers. Tim's wit, candor, and insight transform this from an operator's manual into a good read. His reputation as a collaborative and seasoned leader jumps off the pages here."

Mark Newton, Vice President, Corporate Social Responsibility, The Timberland Company

"Tim really delivers on the promise of the book: namely, a practical guide to making a positive social or environmental impact while working inside a company. He has expertly identified many of the key strategies for success in that kind of role as well as the potential missteps that can be taken. The bonus is that Tim delivers all of this information in a thoroughly engaging manner that conveys his passion for the subject."

Bruce Klafter, CSO AMAT

"*Changing Business from the Inside Out* is the essential how-to handbook for anyone working in, or aspiring to work in, corporate responsibility. Mohin shares insights, inspiration, and invaluable lessons learned from his 20-year career on the front lines of this evolving field. Many sustainability books tell us *why* business should be more sustainable; happily, this one tells us *how*."

Katie Kross, author, *Profession and Purpose: A Resource Guide for MBA Careers in Sustainability*

Transformational change can happen anytime and it can start anywhere. Tim Mohin's *Changing Business from the Inside Out* is a practical handbook for leaders who want to make change happen in small and large ways at work. It is a unique guide seasoned by experience, informed by proven expertise, and inspired by real results.

Mark Pinsky, President & CEO, Opportunity Finance Network

Never one to shy from a challenge, Tim Mohin captures the essence of the opportunity each of us has to make a tangible, measurable, impactful difference working as part of a for-profit enterprise. With this book, there is no longer a need to repeat the learning curve many of us traced in setting the table for our successors.

Find the right company. Take the initiative. Strike the balance between permission and forgiveness. Let Tim show you the way to live your values while working from inside.

Mark Spears, Director, Sustainable Business Practices, Disney Consumer Products; Lead Corporate Director, The Sustainability Consortium Board of Directors

***Changing Business from the Inside Out* provides a unique and much-needed perspective and guide that is useful for both people looking for corporate responsibility (CR) career advice and more experienced managers trying to start their own CR program. Having come of age in the CR field in the same era as Tim, I have had many, many people reach out for CR career advice over the years asking questions about how I got into the CR field and what skills they will need to be successful. *Changing Business from the Inside Out* nails these questions and goes on to provide an in-depth guide for developing a CR program. I wish his book had been available when I was building my program and will definitely refer people calling for career and CR field advice to this book.**

Mark Heintz, Director – Corporate Responsibility & Sustainability, Deckers Outdoor Corporation

Ever-younger generations of leaders are changing the way society operates. Business and society are inextricably linked. As society changes, so must business. *Changing Business from the Inside Out* will guide the business leaders of today and tomorrow to work with companies in order to effect great and lasting change. If you are or aspire to become a leader of corporations I urge you to buy, read, and use this book!

Eric Lowitt, author of *The Future of Value*

"A thoughtfully written roadmap for the beginner or expert corporate responsibility practitioner."

Tod Arbogast, Vice President, Sustainability and Corporate Responsibility, Avon Products, Inc.

"The ability to 'lead from within' is a prerequisite for success in today's business environment. From the new employee to the C-suite, Tim Mohin has written a great guidebook for those working to drive sustainability from within their company."

Dave Stangis, Vice President, Corporate Social Responsibility, Campbell Soup Company

"Finally, a book from the practitioner who has been inside the trenches for 20+ years. A valuable resource for any sustainability professional. A valuable resource for any jobseeker seeking to get into the sustainability profession. Tim has been in the trenches long enough to fill a book with stories. He uncovers the challenges that a chief sustainability officer faces."

Ellen Weinreb, CEO, WEINREB Group Sustainability Recruiting

"A hands-on guide for anyone looking to align their work with their values. *This* is how you grow and differentiate your career, deliver impact at work, and create positive change the world."

Jo Mackness, Executive Director, Center for Responsible Business, University of California Berkeley-Haas School of Business

"Tim Mohin uses his personal career experiences as 'teaching moments' to illustrate so powerfully the many 'lessons learned' throughout the book. This book brings new insights and provides exceptional value for both those seeking a career in corporate responsibility, and for established CR practitioners. *Changing Business from the Inside Out* offers practical, proven methods for building support, dealing with resistance, adapting to change, and manifesting strong leadership skills that are so essential for success."

Clifford Bast, Managing Director for Bast SUSTAINGROUP LLC; former Global Leader of Hewlett-Packard's Environmental/Product Stewardship/Sustainability program

"This book serves not only as a stimulating guide for those entering the field of CSR, but should be read by corporate leaders. As corporations continue to have a larger impact on society, it is an excellent tutorial for decision makers. Managing the triple bottom line is not that difficult and should be embraced rather than ignored."

Bill Sheppard, Former Vice President and Director of Corporate Services, Intel Corporation

"Long before corporate social responsibility was in vogue, Tim Mohin was using his experiences and influence within companies to achieve the greater good. In *Changing Business from the Inside Out*, he shares his philosophical and practical advice with individuals and companies alike to help you leave your own positive mark on the world."

Allyson Peerman, Corporate Vice President, Public Affairs, AMD

"Anyone who has met the Mohin family or worked with Tim would expect that this is the kind of book that Tim Mohin would write. In *Changing Business from the Inside Out*, Tim shows how to run a business with the same values that you aspire to see in your own family and which communities expect of the corporations that provide employment and economic value. This book is about building and supporting successful businesses that make a lasting and positive impression on the world around them and that recognize that business must value more than the single bottom line."

Susan Mac Cormac, Chair of Business Department and Cleantech Group, Morrison & Foerster (2011 California Lawyer of the Year); and Andy Taylor, Senior Consultant, Energy & Environmental Economics

"After spending years writing and researching about careers in corporate social responsibility and sustainability, I've come to realize that anyone with passion, ethical judgment, and an understanding of her environment can be a CSR professional. In fact, there shouldn't be one CSR executive in every company but dozens.

However, becoming a CSR practitioner takes a certain level of understanding. That's where Tim's book fills in the gap for aspiring professionals and jobseekers who want to use their careers to make a difference. There are few executives like Tim who understand the reality of working within corporate corridors knowing that the going will be tough yet continue to persevere because they know it's the right thing to do. There is much to be learned from this book for its emphasis on critical thinking and practical sense. It should be a compulsory handbook for graduating college seniors who must understand the consequences of their future actions."

Aman Singh, CSR Journalist and Editorial Director, *CSRwire*

An inspiring, insightful and insider's under-the-cover view of corporate social responsibility and sustainability. Tim Mohin shares a wealth of practical tips and approaches grounded in over 20 years of front-line experience in some of the most demanding of corporate settings. A must-read book for those looking to embark on a career in this arena, or reference guide for existing practitioners. Tim's book not only articulates the compelling business case for companies to excel in this area, but lays out thoughtful guidance on a framework of how-to actions and key considerations to weigh for long-term success.

Alex Heard, Vice President Environmental Health and Safety, First Solar

Tim Mohin is the Director of Corporate Responsibility at Advanced Micro Devices (AMD). He is responsible for the company's overall corporate responsibility strategy, performance and communications. Prior to joining AMD, Tim advised executives at Fortune 500 companies as the lead sustainability consultant for EORM.

Formerly, Tim led Apple's Supplier Responsibility program. Tim also had a 12-year career with Intel Corporation where he held several positions including Director of Sustainable Development, and Global Environmental Manager. Before joining Intel, Tim worked for ten years in the United States federal government with both the U.S. Senate and the U.S. EPA where he worked on the 1990 amendments to the Clean Air Act and other environmental laws and regulations.

Tim has a bachelor's degree in Environmental Biology from the State University of New York at Cortland and a master's degree in Environmental Management from Duke University.

Read Tim's blog posts at www.timmohin.com and follow Tim on Twitter: @corp_treehugger.

CHANGING BUSINESS FROM THE INSIDE OUT

A Treehugger's Guide to Working in Corporations

Timothy J. Mohin

Routledge
Taylor & Francis Group

LONDON AND NEW YORK

BK

Berrett–Koehler Publishers, Inc.
San Francisco
a BK Business book

First published 2012 by Greenleaf Publishing Limited

Published 2017 by Routledge
2 Park Square, Milton Park, Abingdon, Oxon OX14 4RN
711 Third Avenue, New York, NY 10017, USA

Routledge is an imprint of the Taylor & Francis Group, an informa business

Cataloging information is available from the Library of Congress.
    ISBN-13: 978-1-60994-640-1

British Library Cataloguing in Publication Data:
    A catalogue record for this book is available from the British Library.
    ISBN-13: 978-1-906093-70-9 (pbk)

Cover by LaliAbril.com

Contents

Preface

Like jumbo shrimp or military intelligence, corporate responsibility is considered an oxymoron by much of society. Corporations are among the least trusted of our institutions. As I type these words, anti-corporate sentiment has boiled over, prompting a legion of young people to protest in front of the New York Stock Exchange and across the country in the "Occupy Wall Street" movement. Confirming popular anti-corporate opinion, a global stock trader interviewed by the BBC in September 2011 summed up the current European economic crisis this way: "I go to bed every night and dream of another recession. It is an opportunity for me to make money. Governments don't rule the world, Goldman Sachs rules the world."

It's no wonder that people are skeptical of corporations. The 2008 mortgage meltdown left millions of people in economic ruin. From the BP oil spill in the Gulf of Mexico, to the collapse of Enron, big companies have acted recklessly and the cost to repair the damage has been borne by society in the form of taxpayer-funded bailouts and environmental cleanup. Indeed, as the number and scale of corporate misdeeds mount, it is increasingly clear that governments are incapable or unwilling to protect the public's interest against corporate misbehavior.

With this backdrop, what is the solution? Should policymakers try harder to rein in companies? Should we move away from a capitalist economy and toward a socialist system? Winston Churchill said that "democracy is the worst form of government except all the others." Can the same be said about market-based economies? Without question, corporate greed and negligence have been a source of misery, but the free-market economy has also created abundance and wealth for more people than at any time in human history. Regardless of your worldview on the benefits or ills of capitalism, it is the system we have.

So, at a time when trust in corporations has reached an all-time low, why is interest in corporate responsibility at an all-time high? Skeptics may conclude that corporate responsibility is merely a smokescreen to mask misdeeds. A more plausible explanation is that increasing numbers of stakeholders are demanding responsibility from corporations. Hyper-transparency of corporate activities, fueled by disclosure laws and the Internet, has increased awareness to the point where corporate behavior is under constant scrutiny. Smart business leaders are aware of this scrutiny and of the high costs of a public scandal. They know that in the long run it is cheaper to act responsibly now than to dig out from a PR disaster later.

But there is a more human side of the story. Having spent most of my career working in large corporations, the simple reality is that companies are just groups of people that make very human judgments. Like any group of human beings, each company takes on a unique culture that can either promote ethical behavior or encourage cutting corners. Without question, business leaders are a very competitive lot, but my belief is that most are moral and ethical people.

In his book, *How Good People Make Tough Choices*, Rushworth Kidder defines ethics as "obedience to the unenforceable." With this elegant phrase, Kidder has captured the essence of corporate responsibility: how a business acts when there are no laws or rules to govern its behavior. These decisions are powerful inflection points. The reason I have dedicated my career to corporate responsibility is that by working within large companies, a treehugger like me can steer these decisions toward social and environmental good. And, like steering a supertanker, sometimes a very small nudge in the right direction can produce massive change.

This book is a manual on how to steer the corporate supertanker toward doing good for people and our planet. While being a professional altruist in a for-profit company is a bit like being the designated driver at a cocktail party, it can also be very, very rewarding. There are many examples I could pull from my career, but one of the most touching came from my time as Apple's head of supplier social responsibility. After years of work and millions invested, I could see that conditions had improved for thousands of workers. The most memorable moment was when I walked into a classroom we had set up in the factory to allow the workers to take online courses after their shifts on the manufacturing line. Hundreds of the young Chinese workers used the classroom to learn various topics, and most chose to learn English. When I entered that classroom, the students/workers mobbed me with sentiments of thanks spoken with their newly acquired language skills. In any language, their genuine gratitude for the chance to learn a skill that could improve their lives came through loud and clear.

Whether your worldview is that corporations are inherently selfish or are more prone to act in the public's interest, it is undeniable that the free-market economy is the dominant social institution of our time. The pessimists forecast a race to the bottom where multinational corporations diminish social and environmental conditions. The optimists see an upward spiral of responsible companies working to improve conditions, even making a profit in the process. Whichever view is correct is an abstract academic argument. The reality is that the corporate responsibility movement is real and expanding at a rapid rate throughout the world economy. I wrote this book to help others who feel, as I do, that working in corporate responsibility is the most effective way to make a difference in the world.

Steve Jobs, who passed away recently, was eulogized as an innovator who changed our lives. One lesson from his iconic life that is

applicable to a career in corporate responsibility is to know what
you want to achieve and never compromise on your goals. Almost
twenty years before his death, Jobs summed up his legacy this way:
"Being the richest man in the cemetery doesn't matter to me . . .
Going to bed at night saying we've done something wonderful . . .
that's what matters to me" (Steve Jobs, *Wall Street Journal*, May 25,
1993).

Acknowledgments

This book was a labor of love both literally and figuratively. I wrote it because I love what I do and based on the belief that my experiences in the business world could be helpful to an army of younger people who will become corporate leaders and help to save our planet. The fact is that I could not have had the opportunities that fuel the words on these pages without the support and love of my wonderful family. My wife Catherine and my children Theresa and Jacob are my life.

Special thanks go out to Catherine for both editing this book and putting up with me as I holed up in my office over many nights and weekends. Thanks to Theresa, my daughter and newly minted lawyer, who reviewed her Dad's work with a critical eye. And thanks to Jacob, my son and a PhD student in chemistry, who will someday invent the technology that will keep us all safe.

I also want to thank Mica Odom for her meticulous edits. Mica is an intelligent and energetic young lady whom we are privileged to know. With her MBA from Columbia, she could choose any career but is guided by her altruism. Thanks to Brad Bennett and Katie Kross who both provided excellent perspective and insights.

Finally, thanks to Taimur Burki, who may not even know that he inspired the title of this book on a ride to the airport.

Introduction
Working for good inside a corporation

I was wise enough to never grow up while fooling most people into believing I had (Margaret Mead).

So, you want to save the world, but still need to earn a decent living? If this sounds like you, you have opened the right book.

The question is: how? Corporate jobs aren't likely to send you to underserved communities to teach, and nonprofit jobs don't usually pay very well. Most people entering the job market today are saddled with student loans and are looking for a role that will give them a financial foothold in life. Is it always one or the other – pursuing a living or pursuing your dreams? Do you have to abandon your values to earn a good salary?

The answer is: no. The emerging field of "corporate responsibility" (CR)[1] is an attractive option that spans the traditional border between for-profit capitalism and applying your skills to help people and the planet. This field offers a way to have your cake and eat it too. In other words, you can realize your altruistic goals and still earn a decent living in the corporate world.

But *wait*, you say, aren't jobs in CR rare and hard to get? Yes, this is a new field and, while there are not as many opportunities in CR as there are in more traditional business roles, it is a rapidly expanding area and new jobs are being created all the time. Further, as will be discussed throughout this book, there are many ways to contribute to social and environmental causes outside of the formal CR department. Most companies have a small CR staff that is focused on marketing their CR story, but it is the traditional business functions that *create* that story.

Isn't working for a company selling out?

To some readers, the very notion of working within a corporation is tantamount to selling out their values as advocates for social or environmental justice. While this is a valid perspective, there is another view. Liz Maw, executive director of the MBAs for Social Justice Group "Net Impact," articulated this view in her opening remarks for the 2011 Net Impact conference, when she said, "We are here to occupy Wall Street from the inside." The standing ovation was spontaneous, sustained, and genuine. The audience represented a whole new generation of young people moving into the workforce with their sights set on working for societal good from within a company.

But, as the occupy protests drag on, the popular view is far more divided. Are all corporations greedy and self-interested? Can corporations really be a force for good? These are questions that have been pondered for some time. The legendary economist Milton Friedman authored a *New York Times* op-ed in 1970 titled "The Social Responsibility of Business is to Increase its Profits." Friedman pulled no punches in the opening to this piece:

> The businessmen believe that they are defending free enterprise when they declaim that business is not concerned "merely" with profit but also with promoting desirable "social" ends; that business has a "social conscience" and takes seriously its responsibilities for providing employment, eliminating discrimination, avoiding pollution and whatever else may be the catchwords of the contemporary

crop of reformers. In fact they are – or would be if they
or anyone else took them seriously – preaching pure and
unadulterated socialism. Businessmen who talk this way
are unwitting puppets of the intellectual forces that have
been undermining the basis of a free society these past
decades.

Is corporate social responsibility "undermining the basis of a free
society?" Should companies have any role in protecting people and
the planet? Should the excesses or externalities that can result from
the pursuit of profit be the sole province of government and/or civil
society to monitor and regulate? Friedman and a line of followers (see
"The Case against CSR," *Wall Street Journal* op-ed, 2010[2]) have articulated the popular perspective that companies have no obligation to
people and the planet. Their only obligation to the world is to generate
profits for their shareholders.

BUNK!

Such black and white distinctions only make sense in the academic
ivory tower. In the shades of gray that color the real world, companies
must make trade-offs every day on where to invest and how to conduct their business. High-profile cases of corporate misconduct mask
the less sexy, but no less important, cases of companies choosing to
do the right things right. Even Friedman admits that business leaders
must conform to the basic rules of society and ethical norms in his
1970 article:

> [The] responsibility [of the business executive] is to con-
> duct the business in accordance with [the owners'] desires,
> which generally will be to make as much money as pos-
> sible *while conforming to the basic rules of the society,
> both those embodied in law and those embodied in ethical
> custom* (emphasis added).

The "ethical customs" of society have changed a bit since 1970 when
this article was published. On December 2nd of that year, President
Richard Nixon created the Environmental Protection Agency (EPA) in
response to public outcry over corporate pollution disasters such as
the near extinction of songbirds from the use of the insecticide DDT

(unveiled by Rachel Carson's 1962 book *A Silent Spring*), the Cuya-
hoga river fire in 1969 (yep, the river actually caught on fire), and the
first Earth Day held in April, 1970.

So, by following the "ethical customs" before 1970, rivers caught
on fire and songbirds were driven to the brink of extinction. Thank
goodness today's ethical norms are more enlightened. Society expects
more from corporations and, as these expectations increase, there is a
growing need for people to work for social and environmental justice
from inside companies.

By effectively working within a company you can influence decisions that can have massive societal benefits across the globe. And
there has never been a better time to work on these changes. The race
to be the greenest, most responsible company on the planet is under
way (e.g., last year, more than 5,500 companies around the world
issued sustainability reports,[3] up from about 800 ten years ago) and
appears to have substantial staying power. Companies of all types are
looking for people to help improve their environmental, social, and
ethical performance. By learning the skills and strategies of working
for good within a company you can create large, immediate, and lasting change.

Instead of empty rhetoric, this point of view is the essence of my
own career choices. I have done more for people and the planet working within corporations than I could have ever expected to achieve
had I stayed in the government (I worked at the U.S. Environmental
Protection Agency and the U.S. Senate in the first ten years of my
career). While government regulators and nonprofit activists are very
important drivers for social and environmental justice, they must
work from the outside to cajole companies into good corporate behavior. The threat of enforcement or activism as a tool for change pales in
comparison to the sweeping implications of, for example, leveraging a
multinational corporation's buying power to transform working conditions in a global supply chain.

To a certain extent, being a corporate treehugger is a line-walking
exercise. Corporations are indeed focused on profit, and being an activist within a company is very different than being an activist for a nonprofit organization. But as expectations and transparency increase, the

"ethical customs" for corporate behavior are changing. These macro-level changes are opening up new jobs in CSR and changing "main-stream" roles across almost all corporate functions.

I wonder if Milton Friedman would think that the inmates had taken over the asylum if he could witness 2,600 enthusiastic MBA students and professionals cheering for corporate responsibility at the 2011 Net Impact conference. As these business leaders of tomorrow increasingly occupy Wall Street from the inside, even Friedman might have to concede that the profit motive and social justice can be mutually supportive.

What you will get from this book

The idea behind this book is to provide practical advice for people looking to enter the world of CR, either in the official CR department or within a more mainstream role. This book is not about theories, case studies, or abstract business strategies. There are numerous books presenting hundreds of theories about how companies can both "do good and do well" through corporate responsibility. While these theories are important, they are often not practical unless you are the CEO or have similar decision-making authority. For the rest of us – those who haven't made it to the executive suite just yet – this book is a how-to manual for contributing to social and environmental well-being through a career in business.

The recent explosion of interest in CR has created an exciting new career path and new job opportunities to work within a for-profit company while pursuing altruistic goals. For those who are interested in the CR field, this book outlines step-by-step tips for designing and running a successful program as well as the essential skills and attributes for this career path. For those who are not interested in working in the CR department, the guidance in this book can be applied from almost any position within a company. As you will see, the opportunities to contribute to society may be even greater from outside the CR department. The key is to *lead from where you stand.* Anyone in any department and at any level can make a difference.

I wrote this book because I wanted to share the practical lessons gained over more than 25 years of work that included much trial and error. As this field has grown, it has also become popular with a legion of young adults who are looking for their first job as well as career-switchers who want more meaning from their work. This trend is inspiring and motivated me to share what I have learned. My intent is that this book provides some practical guidance for those who seek a career in corporate responsibility. While it is obvious that not every tip will work in every situation, I hope that the stories and examples in the pages that follow will inspire you to make career choices that will make a difference.

Why corporate responsibility?

There are many reasons why increasing numbers of people are interested in the emerging field of corporate responsibility. As the reach and resources of corporations have increased, so has their ability to drive meaningful improvements. This broader reach, coupled with heightened awareness and scrutiny of corporate operations, has led to the emergence of corporate responsibility as a viable and meaningful career choice. Specifically, I believe there are three reasons for the rise of corporate responsibility:

1. Business is the dominant social institution of our time

In a globalized economy, the revenue of multinational corporations dwarfs the gross domestic product (GDP) of some countries. Many companies are now large enough to affect change on a global scale; their physical impacts and policies transcend national borders and the decisions made in corporate boardrooms can help or harm millions of people.

Take Walmart, for example. Love them or hate them, as of this writing, Walmart is now the world's largest company with revenues exceeding US$400 billion and approximately 2 million employees (or associates, as Walmart prefers to call them). Mentioning Walmart in the same sentence as corporate responsibility will elicit strong reactions in some circles. Many people associate the brand with everything that is wrong with corporate America – from poor wages and benefits for their employees to shutting down small businesses by undercutting their prices. However, over the last several years, Walmart has developed an impressive sustainability program. In 2009, Walmart introduced its Supplier Sustainability Assessment Survey – a diabolically simple set of 15 questions examining the actions of its suppliers in order to better protect the environment and uphold labor rights. The survey is diabolical because to answer affirmatively, each question requires the supplier to have fairly sophisticated CR programs. For example, question one asks: "Have you measured and taken steps to reduce your corporate greenhouse gas emissions (Y/N)?" This Yes/No question sounds straightforward, but to answer "Yes" requires an in-depth carbon footprint[4] study that many companies have yet to undertake. With the scale of Walmart's purchasing power and the not-so-subtle commercial pressure to score well, this simple survey has already had wide-ranging impacts for manufacturing companies around the world. To up the ante, Walmart has publicly discussed its intention of turning the scores from this survey into a point-of-sale sustainability ratings label.

The scale of Walmart's turnaround is hard to overestimate. Jeffrey Hollander, the co-founder of the cleaning product eco-brand Seventh Generation, was quoted as saying, "Hell would freeze over before Seventh Generation would ever do business with Walmart." But in a 2010 interview with FastCompany.com he said, "They aren't the same company they were when I said what I said. I'm the first one to admit that I was naive in thinking it was impossible for them to change."

As the Walmart example demonstrates, working for social and environmental improvement within a large company can be effective on a massive scale. As we will explore in this book, it can also be monumentally frustrating. Companies are profit-making institutions beholden to return maximum profits to their shareholders. In the past, this meant that companies cut corners to save money whenever it was legal and/or expedient. The good news is that today this kind of behavior is increasingly unacceptable. Watchdog groups monitor company behavior closely and have become very adept at drawing attention to corporate misdeeds through "name and shame" or "rank and spank" tactics. New laws such as Sarbanes–Oxley[5] and Dodd–Frank[6] continue to push companies into greater levels of transparency and accountability. This increased scrutiny and the growing awareness and expectations from customers, employees, and the general public have made it imperative for most companies to build a strong CR program.

2. Opportunity

Strange as it may seem, there is money in altruism. As the world runs short of resources there is an increasing market for more efficient products and services. The economic winners of tomorrow will be the innovators who find ways to do more with less – to stretch our finite resources. Continuing to prosper while using fewer resources is the definition and aspiration of sustainable development.[7]

While there is a strong argument that the excesses of industrialization have led to many of the social and environmental problems we now face, there is an equally strong point that business will also produce the solutions. Business – or, more accurately, the profit motive – represents an incentive system that rewards the iconoclasts and creative innovators who can see around corners, think differently, and are willing to take risks. In today's world, these inherent business incentives are increasingly being applied to help people and the planet. For example, rising fuel prices and the threat of global warming have shifted the auto industry away from making gas-guzzlers, and toward innovating with higher-mileage hybrids and electric vehicles. General Electric (no one's idea of a corporate treehugger) has claimed billions in profit from its Ecomagination™ campaign – designing, marketing, and selling technology based on resource efficiency.

Today's workforce increasingly looks at the world as a set of problems to solve rather than just a set of markets to exploit. Some may see this as a distinction without a difference, but the difference is profound. World population has doubled since 1960 and will double again in the next 50 years. With the rise of the emerging economies in China, India, Brazil, and others, our well-being and survival depend on the sustainable use of natural resources. Whether you subscribe to Adam Smith's *Invisible Hand* economic theory (i.e., price signals control everything),[8] or believe that government regulation is the only effective way to control the excesses and externalities of business, eventually it does not matter – in a resource-constrained world, sustainable innovation will always win.

"Build a better mouse-trap and the world will beat a path to your door" is the widely quoted phrase attributed to Ralph Waldo Emerson as a metaphor for the success that springs from innovation. Combine this axiom with the ancient proverb "necessity is the mother of invention" and you have the business case for sustainability.

The megatrends of increasing population, dwindling resources, and increasing pollution will spawn the major industrial powers of the future. These new super-companies will be global businesses led by bright young minds that will discover new ways for humanity to thrive without using up the planet on which we depend.

3. Legacy

The third major factor leading to the rise of corporate responsibility is the desire to leave a positive legacy. Clayton Christensen, author of *The Innovators' Dilemma* and a guru in the business world, summed up this drive brilliantly in an article titled "How Will You Measure Your Life?"[9] The article drew attention, not because it was another of his phenomenal observations on business strategy, but because it was a very personal exploration of his life's legacy. Christensen makes the case that following the business law of maximizing marginal returns can lead to ruin. He distinguishes his career from his Harvard Business School classmates who initially found riches but ended with a life in tatters. Guided by his values, Christensen eschewed the traditional business doctrine, instead investing his considerable talents in research, teaching, and family. The take-home message of this paper is that there are far more relevant and important ways to measure your life than in dollars.

Another great example of a value-based business career is Gary Hirshberg, the former CEO of Stonyfield Farm (an organic yogurt company in New Hampshire). Rather than compromise his values, he founded a successful company based on sustainability and social equity. In his speech to young MBA students at the 2010 Net Impact conference, he outlined how he struggled through the early years after starting his business. He described how the experts derided his business model because organic foods were more costly and harder to produce. To ensure an adequate supply, he worked with farmers to teach them how to produce organic food and showed them that they could make more money with less price volatility. When the company gained enough of a foothold in an industry dominated by giant food conglomerates, he answered their multi-million-dollar advertising attacks with innovative packaging that contained quirky messages, yogurt container take-back programs and "guerilla" advertising tactics (i.e., people-to-people) to draw attention to the brand. He proudly reported that his struggling organic foods start-up is now commanding industry-leading margins and growth rates while continuing to pay dairy farmers higher prices. At the end of his remarks, Hirshberg left his most enduring mark when he declared: "You should never compromise your values for work. If a company makes you check your values at the door, find somewhere else to work."

This profound statement generated a spontaneous and sustained standing ovation from the packed house and, I would guess, spawned more than a handful of new sustainable business start-ups.

Hirshberg's enviable legacy shows us how to hold on to your social and environmental values, make your fortune in business, and redefine an entire industry. Christensen's example shows us how to live a values-based life and define your legacy with accomplishments, not just dollars. Both men are icons for the central theme of this book: *how to make your living AND make a difference.*

My journey

I started working at the United States Environmental Protection Agency (EPA) right out of graduate school at Duke University's environmental program. The terms "sustainability," "corporate responsibility," and "corporate social responsibility" had not been invented yet. My choice to work in environmental protection was a reflection of my personal values and was considered a bit eccentric at the time.

After self-funding my education by taking on substantial debt and choosing a major with limited economic prospects, I faced significant pressure to find work. When I graduated, the clock began ticking on my student loan payments that would come due in six months.

My résumé had been circulated to the EPA's offices in Durham, North Carolina, where its national program offices for air quality are located. At the time, EPA was under pressure after some highly publicized scandals resulted in the resignation of the administrator.

Bill Ruckelshaus – who had been EPA's first administrator – had just come back to lead the Agency and, in Congressional testimony, committed to 20 new regulations of toxic air pollutants in just one year. To put this commitment into perspective, in the 15-year history of the Clean Air Act (at the time), only seven toxic air pollutants had ever been regulated. Ruckelshaus essentially wanted to triple this number in one year.

My field of study at Duke (environmental toxicology) matched the skills EPA needed to fulfill this commitment. After a few nail-biting months, I was hired into an entry-level role. I would love to claim that my career choice after graduate school was based on my values, but had I been trained in anything else, economic necessity might have led me down a different path.

A few years later, the overhaul of the Clean Air Act became a campaign promise for presidential nominee George H.W. Bush. Early in his administration, EPA was asked to draft the president's version of amendments to the Clean Air Act. I was selected to be on the drafting team for the toxics section. The next two years were spent shepherding the bill through the Congressional process, culminating with a signing ceremony in the East Room of the White House on November 15, 1990.

I returned to more standard EPA work – writing rules – for the next couple of years, and then took a fellowship to work on Capitol Hill. The fellowship program "loaned out" EPA staff to Congress and I landed a job working for Senator Max Baucus, a Democrat from Montana, who had recently taken over as Chairman of the Senate Environment and Public Works Committee.

Over the next year I worked on drafting all sorts of environmental legislation including a bill promoting environmental technology that was Senator Baucus's signature policy (it passed in the Senate 85-15 but never became law). As my fellowship year drew to a close, Senator Baucus decided to keep me on his committee staff.

Only one year later, in the elections of 1994, Republicans swept in to majorities in the House and Senate. Senator Baucus lost his committee chairmanship and had to reduce his staff by half. After a few nervous weeks, Senator Baucus offered me a role to stay on at Capitol Hill as a member of his minority staff.

During this time, I had heard about a job with Intel Corporation and decided to check it out. Intel's values and objectives fitted me well. They needed someone in Washington, DC to help them reduce the bureaucracy that might slow their unprecedented growth but, to a person, everyone I spoke to at the company cared deeply about the environment. I decided to leave the Senate and became Intel's first government affairs manager entirely focused on environmental policy.

After three years in the government affairs office, an opportunity opened up to run the environmental department for Intel globally. Intel's director of environment, health, and safety recruited me for the role. I took the job, and in January of 1998, I transitioned from a policy-focused career to a career focused on improving the global environmental performance of a major multinational corporation.

This is where my story – how to be a treehugger in the corporate world – begins. While I had been working in the environmental field for more than ten years, there were many lessons that I was to learn about working inside a big corporation which make up the pages of this book.

In addition to nearly 12 years at Intel Corporation, I have since worked at Apple, where I started and led the supplier responsibility program focused on Asian factories making iPods, iPhones, and Macintosh computers. After a brief stint consulting to a number of companies on sustainability, I am currently working as the director of corporate responsibility at Advanced Micro Devices (AMD). These roles, each challenging in their own way, have provided me with insights into the tactical, day-to-day duties of a corporate responsibility manager.

Using this book

This book covers the knowledge, skills, and abilities that you will need to land a job and be successful in the growing field of corporate responsibility. Like all business books, the lessons here are situational, meaning that they will be useful in some situations but not in others. Each chapter focuses on an important skill area or substantive topic that can be readily applied in the appropriate situation.

Every chapter begins with a summary of the main ideas covered, gives an introduction to the topic and then outlines the basic skills, practical tips, and a few anecdotes that are helpful for managing the issue. The breadth of corporate responsibility makes it impractical to become an expert in each of the topic areas; thus, the chapters deliberately deliver an overview as opposed to an in-depth treatise.

In addition to the practical tips and stories on how to manage the various technical components of a corporate responsibility program, this book provides information on the "softer skills" you will need for success in corporate responsibility. These skills include communications, reading the system, leading through influence, and others, which can be applied to a broad range of careers beyond the world of corporate responsibility.

The name game

By now you may have noticed that the terms "sustainability," "corporate responsibility," and "corporate social responsibility" are used synonymously. There is a great deal of confusion around the definition of these terms within this profession. Many companies are now adopting the title "corporate responsibility" because it encompasses environmental, social, ethics, and governance aspects of the job. I will default to the more inclusive "corporate responsibility" for this book, but these terms are often used interchangeably.

The term "treehugger" in the title and throughout the book sometimes elicits the stereotype of an environmental zealot. While this is a valid perspective, my hope is that most readers will not be turned off or feel that this is an overly limiting term. While "treehugger" is often associated with an environmental focus, I use this term to encompass a wide range of altruistic viewpoints. In the end, I hope to provide practical guidance to help you accomplish altruistic goals through a corporate career.

Lead from wherever you stand

This book is full of advice on how to set up and run a successful corporate responsibility program. If you find yourself in a job where you have limited authority to use these lessons because the role is constrained to a narrow scope, there are still ways to achieve your goals. Never let your position supersede your passion or overshadow your abilities.

All great leaders started from lower-level positions and discovered ways to leverage their capabilities to add value to their organization. You can use the ideas in this book to contribute to the well-being of people and protect our planet, regardless of your position in the organization and, in doing so, enhance your career and your legacy.

1

The department
of good works

This chapter covers the evolution of the corporate responsibility function within companies, its organizational structure, and how these factors affect the practice.

Doing well is the result of doing good. That's what capitalism is all about (Ralph Waldo Emerson).

I believe that most people would claim that they want to do good things for others. But, when it comes to their careers, that inner altruism is often forced into the back seat. At the risk of stating the obvious, "the business of business is business." When you go to work for a company, whatever your role may be, you are expected to deliver a return on the company's investment in your pay and benefits. For most roles within big companies, this means that saving the world may not fit into your job description – at least at first glance.

The reality of working in a big company is this: you are given a set of goals, your progress toward those goals is measured (usually in periodic performance reviews), and your performance is typically ranked against your peers (which is somewhat counterintuitive since most companies extol the value of teamwork). The environment is competitive, with money and advancement up the corporate ladder on the line. On top of the intrinsic competition, and lurking in the background at most companies, is the very real threat of layoffs that are increasingly a fact of life. All of these pressures drive you to keep your head down and remain focused on delivering on your goals, not taking time out to help make your company more sustainable.

Most of the books on sustainable business give you academic models, case studies, and high-level theories on how to embed sustainability into business or how companies can unlock hidden value by investing in corporate responsibility. As discussed in the introduction to this book, all but the most senior of corporate jobs don't have the scope or authority needed to drive this type of change. This book focuses on practical, proven strategies to drive sustainability from a staff-level position.

To set the context, this chapter describes the genesis of corporate treehugger jobs. Corporations have had small groups of people in altruistic roles for many years. Recently, the number of these jobs has been increasing along with rising awareness and importance of corporate responsibility.[10] This chapter lays out the evolution and *raison d'être* of business functions that contribute to society, from the traditional community affairs team, to a full-fledged corporate responsibility department.

The evolution of corporate responsibility

Steven Covey, in his timeless classic *The Seven Habits of Highly Effective People*, coined the phrase "seek first to understand" (Habit 5).[11] Following this rule, let's first step back and reflect on how social responsibility became intertwined with business. Starting in the 1970s, as the fundamental environmental protection laws were put into place, companies were on one side and environmental groups were on the other side while the government played referee and set the rules and regulations. As these battles subsided, something happened: the environmental movement won the hearts and minds of the public. With public support for environmentalism polling overwhelmingly favorable, companies changed their tactics from fighting regulation toward voluntarily "greening" their operations.[12] Since that time, the traditional interest groups have been rearranged. Instead of lining up against environmental regulations, most of the Fortune 500 companies are now spending significant resources to demonstrate their sustainability credentials.

Like the end of the cold war, the victory of the environmental movement slowly filtered into our collective consciousness and a new world order emerged. In the new world, all things green are good. "Green" has become a verb. Cigarette packages carry environmental messages and even defense contractors publish sustainability reports. Being against "green" in the new world is like being against technology. Raising objections to sustainability will guarantee your status as a Luddite.[13] The future, it seems, will be green (or, for the more cynical, at least be branded with green leaves printed on Forest Stewardship Council certified post-consumer paper with soy-based ink).

The point is that most companies are no longer actively (or openly) campaigning against environmental rules. On the contrary, most big companies are now tripping over each other to demonstrate their good works and many are finding value beyond enhancing their brand reputation. To do this work, companies need a **department of good works** – a group of people who are responsible for the care and feeding of the company's reputation as a responsible corporate citizen . . . the tree-huggers in the corporate world.

The battle for the corporate soul

An example of the shift toward corporate "greening" was the near-mutiny at the U.S. Chamber of Commerce in 2009 when PNM Resources Inc., Exelon Corp., PG&E Corp., and Apple resigned from the Chamber over differences on climate policy (Nike also resigned from the Chamber's board of directors but

→

vowed to remain a member company). These companies took a visible and principled stand to oppose the Chamber's regressive stance on pending U.S. legislation to protect the climate.

To borrow a phrase from Malcolm Gladwell, these high-profile defections from the Chamber signaled a "tipping point."[14] It seemed that the companies who left the Chamber had reached the point at which the risk to their brand image for being seen as opposing climate protection legislation was greater than their potential savings from avoiding climate regulation.

One of the themes of the film *The Corporation* is that companies will always act out of their own self-interest. If this is true, the defections from the Chamber might signal that the benefits of a responsible reputation are now far more valuable than avoiding the costs of climate change legislation. Perhaps these defectors were the vanguard of companies making rational, self-interested decisions in favor of the environment because to do so makes economic sense for them. By leaving the Chamber, they stood to benefit by protecting and enhancing their reputations as responsible environmental stewards. From this perspective, distancing themselves from the Chamber's efforts to dilute or defeat the climate bill was a rational business decision.

But this change may have been even larger than it seemed. The Chamber represents its members on a wide variety of legislative issues. According to the *New York Times*, the Chamber's 2010 budget was approximately $200 million.[15] This is a formidable war chest of resources to influence a broad range of legislation beyond the climate bill. So, by exiting the Chamber, these companies were, in effect, saying that the value of their green reputation exceeds the overall value they would get from the Chamber's substantial lobbying might *on all issues*.

This case is about the battle for the corporate soul on environmental issues and corporate responsibility. If the value of being green now exceeds the value of avoiding costly regulations, it represents a sea change in thinking by corporate leaders.

Where corporate responsibility lives

There is no standardized organizational structure for the corporate responsibility department. In many companies the function grew out of the environment, health, and safety department, but it can exist almost anywhere within the corporate structure. There are, however, a few departments and functions that are traditionally focused on corporate responsibility issues:

Public affairs and community relations

A traditional home for treehuggers in a big company is in the public affairs or community relations department. These are usually small groups of people whose role is to build relationships with the local community – such as citizen groups, elected officials, and other influencers. They donate corporate money to worthy charities, field complaints from neighbors, sit on local boards, and coordinate employee volunteers to work on community projects. In larger companies, these departments often manage corporate philanthropy by running the company foundation, which donates money to selected charities and often matches employee giving.

Many companies also combine their government affairs – or lobbying function – into this organization. In essence, the public affairs group is like a public relations firm for the company – working at the interface of the company and the public to boost the company's reputation and influence policies that affect its interests.

Companies have invested in these groups over the years for a number of reasons, all of which can be summed up under the phrase "license to operate." The public affairs team represents the company's story about job creation, economic growth, and prosperity, invests in community programs, and builds the relationships that help the company operate smoothly. This work builds the reservoir of goodwill needed to obtain permits, reduce taxes, or gain favorable policies without picket lines, protests, or costly delays.

Intel vs. Sumitomo: The license to operate

A case study of the "license to operate" paradigm comes from my days at Intel Corporation. In the late 1990s, the company was building a huge computer chip factory in the suburbs of Phoenix. The facility would use loads of water and hazardous chemicals, take up acres of land, and increase noise and traffic in the area. The Intel public affairs team fanned out to woo influential civic leaders. They held stakeholder sessions, listened attentively to concerns and took actions – often expensive actions – to reduce the factory's impacts on the community. They also worked with state officials to secure the most favorable tax and regulatory situation possible.

In the end, the permits were issued, the tax breaks were granted, and the factory was built in one of the more affluent areas of town. The community's reaction was overwhelmingly positive and, even through massive expansions (Intel has since built two more factories on that site), the relationship with the community has remained amicable.

Standing in stunning contrast was another computer chip company, Sumitomo, which had planned to build a semiconductor factory in Phoenix at about the same time. This facility would use a similar process to make semiconductor wafers, only it was supposed to be smaller and would be located in a less affluent part of town which, theoretically, would make it easier to get the needed permits. But Sumitomo didn't invest in the community or do much outreach with neighborhood activists. Instead, it focused its efforts on obtaining favorable incentives from the city. The public hearings over the permits for this facility erupted into a firestorm of public protest and outrage that nearly scuttled the entire facility, delayed construction, and damaged the company's reputation in the community. While the total costs of this incident were not made public, they surely exceeded the amount Intel invested in community outreach for its new factory in the same city.

This case demonstrates the value of community outreach to a company and why most medium-to-large companies employ several people in a public affairs function. In many companies, the public affairs department either grew into or became the organizational host for the modern corporate responsibility function.

Environment, health, and safety

Another traditional hub for treehuggers in the corporate world is the environment, health, and safety (or EHS) department (these groups are referred to as EHS, HSE, or SHE; this book will use EHS). The treehuggers tend to hang out in the "E" part of the EHS team – the environmental group.

The environmental department is where I cut my teeth as a corporate treehugger. My position as global environmental manager at Intel had responsibility for environmental compliance for 17 manufacturing sites (and numerous other facilities) scattered across the planet as well as company-wide environmental sustainability and product environmental standards. My time and attention were consumed by tactical management considerations such as knowing which regulations applied to which facilities, monitoring emissions and resources, and dealing with permits for new facilities. When I started in this role, the term "sustainable development"[16] was still fairly obscure and, like most companies, the primary focus was on compliance obligations.

In these early days, the environmental compliance function had the attention of senior management, and for good reason: noncompliance can cost a company a lot of money. Intel got a wake-up call from some Superfund sites (toxic waste that had filtered into underground water sources) that the company was responsible for cleaning up. The company has grown so large since then that it is hard to imagine this today, but the cleanup was so costly that it had a material impact on the quarterly earnings reports. Then came the Clean Air Act of 1990. Intel viewed the rigorous and restrictive permitting requirements of this new law as a potential deathblow to its ability to build the factories and invent the manufacturing processes at the pace it needed to stay on top in the hyper-competitive semiconductor business.

With stakes this high, the EHS department had plenty of attention from the executive suite. Environmental operations were reviewed every quarter with then CEO and company founder Gordon Moore (author of Moore's law[17]) sitting in to oversee the group's performance. Every permit, cleanup site, and regulation was scrutinized in those meetings.

As the compliance and cleanup issues waned and public awareness of sustainability grew, responsibility fell to the environmental department to develop Intel's sustainability program. The program officially started when Intel issued its first public report in 1995, which was called the "environmental report." This report focused on the environmental improvements that the company accomplished over and above what the regulations required. The company had invested millions of dollars in pollution controls, process changes, and product changes to reduce its pollution and consumption of natural resources. Many of these investments were over and above what was required by the applicable regulations.

There were many internal meetings where Intel struggled over the return on these investments. Typically, my team was able to calculate the costs with precision, but the benefits were harder to quantify. During decision meetings the Intel staff often talked about values – such as "community goodwill" – which are vague and hard to monetize. The Intel management during this time deserves credit for making these investments in the face of unfavorable or unknowable return on investment (ROI). Most of these meetings ended when the decision-maker would ask, "Guys, is this the right thing to do?" We would go around the table and nod in the affirmative, at which point we agreed to move forward.[18]

In many companies, similar to the Intel case, the environmental team was the birthplace of the sustainability program. While the environmental function continues to be a central component of a sustainability program (especially for manufacturing businesses), the expanded scope of issues and expectations from stakeholders has outgrown the capabilities of the environmental department.

The growing gap between EHS and sustainability

A few years ago, I gave a talk to a group of EHS professionals illustrating the growing gulf between environmental compliance and sustainability. Titled "Sustainability 101," I started the session asking for a show of hands on "who can define sustainability?" No hands went up and all I heard was the sound of crickets chirping. This audience was far more focused on the nuts-and-bolts of environmental management than the esoteric definition of sustainable development – the perfect group for my Sustainability 101 lecture! The widely used Brundtland Commission definition of sustainable development is "development that meets the needs of the present without compromising the ability of future generations to meet their own needs." This sounded like so much hot air to this group of compliance engineers.

After the event, it was illuminating to talk with some of the participants to understand the chasm that has grown between the practice of environmental management and the concept of sustainability. Sustainability is based on the "triple bottom line" measuring economic, environmental, and social performance. This broad scope includes issues, such as fair labor standards, diversity, and ethics that are beyond the scope of the environmental compliance function.

But there remain strong connections between the environmental management discipline and sustainability. I challenged this group to leverage their technical skills in the environmental field to be successful in the emerging practice of corporate responsibility/sustainability. Using my own career and stories from colleagues' careers as examples, I focused on the essential skills and capabilities needed to transition from a career in EHS to a career in corporate responsibility (discussed in Chapter 2). The take-home lesson for this audience was that it is easier to start with a technical background and acquire the other skills needed for a career in corporate responsibility than it is for non-technical people to learn what this crowd already knew.

The corporate responsibility department

The corporate responsibility department is an aggregation and integration of environmental, social, ethics, communications, and other disciplines. As discussed in the introduction to this book, companies use several names to describe this function, the most common being: corporate social responsibility, corporate responsibility, corporate citizenship, and sustainability. The term "corporate responsibility" seems to be gaining in popularity because it is the most inclusive.[19] While the roots of this function are often in the EHS or the public affairs department, the practice of corporate responsibility both integrates and expands beyond the traditional boundaries of both of these groups.

The best way to define the scope of the issues covered by the corporate responsibility department is the Global Reporting Initiative (GRI) guidelines.[20] The GRI guidelines have become the de facto international standard for disclosures on corporate responsibility matters. They are a collection of more than 100 performance indicators covering everything from corporate strategy, governance, economic performance, executive compensation, environmental performance, labor issues, human rights, impacts on society, diversity, and product responsibility, among others.

One of the primary responsibilities of the corporate responsibility department is to assemble and report on all of the information called for by the GRI guidelines in a corporate responsibility report. The number of companies issuing corporate responsibility reports (these reports can also have many names such as corporate citizenship, sustainability or corporate social responsibility reports) has grown exponentially over the last decade. The CorporateRegister.com tracks more than 38,000 corporate responsibility reports from 8,703 different companies across 159 countries. According to its website, the number of corporate responsibility reports has grown by 672% between 2000 and 2010.[21]

The role of the corporate responsibility department does not end with the annual CR report. This group must also coordinate, cajole, and motivate the corporate functions that produce the data included in the report. For example, because diversity is an important component of overall corporate responsibility performance, CR managers must work closely with human resources (HR) managers to understand and influence their diversity policies. The same scenario is repeated for just about all of the corporate departments. The role of the corporate responsibility manager is to align, influence, and communicate the performance of a very broad swath of issues that span the corporate structure. As discussed in Chapter 2, it takes a special set of skills to lead a function where you have broad responsibility for issues that you have almost no authority to control.

For the most part, corporate responsibility program managers must "lead through influence" in programs they do not directly control. In some cases, however, corporate responsibility departments "own" certain programs (i.e., they have direct authority). These can be "orphan programs," meaning that no other department is managing the program but, because it is critical to corporate responsibility performance, the CR team takes leadership. For example, if the procurement team does not have the resources, expertise, or inclination to manage "supplier responsibility," the CR team might step in to set up a code of conduct and a compliance program (see Chapters 6 and 7). In other cases, the corporate responsibility team might own programs that are remnants of their historical beginnings. For example, CR teams that grew out of the environmental department might continue to own the management of the company's carbon footprint.

Ultimately, the corporate responsibility job is an eclectic mix of leading through influence, leading through authority, and communicating to the public about a very broad range of issues. To do this job well you have to have an understanding of a wide range of corporate functions, be able to influence them, and be able to communicate about them to diverse audiences. The essential skills required to do this job are covered in Chapter 2. Establishing and running a successful program are the focus of Chapters 3 and 4. Chapters 5 through 13 cover the knowledge about all the various programs under the rubric of corporate responsibility. Because communication skill is such a vital aspect of this career path, Chapters 8 and 9 (spoken and written communication, respectively) are focused on this topic.

While there is little consistency in organizational structure or capabilities of corporate responsibility departments today, as the field grows and matures there are efforts to articulate standards for the practice of corporate responsibility.[22] Over time, there will likely be more consistency in the roles, responsibilities, and placement of the corporate responsibility function within the corporate structure.

Choosing the right company

After working in corporate responsibility roles at Intel, Apple, and AMD, I came to this observation: there are two ways that most companies decide to initiate a corporate responsibility function: the "epiphany" or the "2x4."

Based on my own experience, corporate responsibility usually starts with a whack – the metaphorical 2x4 – upside the head. In these cases, the company is suddenly faced with a significant business risk that must be addressed through corporate responsibility measures. In Intel's case, the 2x4 moment was the threat to its business from increased environmental regulation (Superfund cleanup and Clean Air Act permits). In Apple's case, it was allegations of sweatshop conditions in the factories making its iconic iPod. There are many other examples of the 2x4 effect, such as the highly public revelations of sweatshop conditions in factories supplying Nike and Gap. In my experience, this is the most common way that companies decide to invest in this function. Once one company experiences the 2x4 effect, it serves as a wake-up call to other companies in the same sector. Often, one whack is enough to move a large percentage of the industry out of fear that they will get whacked with the same piece of lumber.

The epiphany scenario is when the founder or CEO of the company declares that his or her company will be a leader in responsibility. Great examples of this scenario are Ray Anderson (the late CEO, Interface Carpets), Gary Hirshberg (former CEO, Stonyfield Farm), Dame Anita Roddick (the late founder of The Body Shop), Jeff Swartz (former CEO, Timberland), Ben Cohen and Jerry Greenfield (the founders of Ben & Jerry's), or Jeffrey Hollender (former CEO, Seventh Generation). Each of these business leaders deliberately set out to integrate corporate responsibility into their business strategies. This is very different paradigm than the 2x4 model because these programs are not born out of crisis management or problem-solving mode. These companies are dedicated to making a profit while *simultaneously* helping people and the planet. Epiphany companies often use their stance on sustainability and social justice as a differentiator in their market. Because the responsibility message is coming from the top, it becomes ingrained into the DNA of these companies, and thus is manifest in just about every corporate function and job.

This is an extremely important distinction for the treehugger looking for work in the corporate world. In epiphany companies, you are likely to find opportunities to work on responsibility issues in any job. In 2x4 companies, the jobs that impact sustainability issues are more limited and changing the company's behavior is more challenging.

Based on this paradigm, you probably want to work for an epiphany company. The problem is that there are far fewer of them and the competition for jobs is stiff since so many new graduates want to work for a values-based company.[23] Don't despair: there are still plenty of opportunities to contribute to social and environmental improvement in 2x4 companies, and since that is where the vast majority of treehugger jobs exist, this book will focus there. While there is little consistency in the structure of corporate responsibility departments in these companies, there is a common set of skills and duties that apply.

Group gravity

The old axiom of "where you stand depends on where you sit" comes into play depending on where your CR department is located within the corporate structure. The gravity of the function and culture of your organization will affect the make-up and focus of the corporate responsibility department. For example, if your CR team sits within the marketing department, you will likely have access to the marketing decision-makers and budget. Naturally, your efforts will lean toward weaving responsibility into brand or product messages. At Apple, I worked in the procurement or "operations" group and, predictably, my team was focused on how we could drive responsibility through supplier relationships. At Intel, my team was in the EHS department, which drove our focus into facilities and compliance issues.

Regardless of the tilt that comes with the organizational structure, the scope of issues that the corporate responsibility group must understand, influence, and communicate extends beyond the boundaries of traditional corporate roles. While this can be frustrating, it can also be exciting and satisfying. Very few jobs, outside of the executive suite, have access to more of the company's functions, and fewer still have the opportunity to influence these functions to improve social and environmental outcomes.

2

Skills for success in corporate responsibility

This chapter outlines the essential skills and personal attributes needed for success in a corporate responsibility career.

Skill is the unified force of experience, intellect and passion (John Ruskin).

While numerous graduate programs are popping up that offer training in sustainable business and corporate responsibility, very few people in the CR field have these degrees. Most people working in CR positions have education and experience in some other area and have followed their passion to get to one of these jobs. After talking to several of my colleagues and thinking through my own experiences, I identified some core skills that are important elements to success in corporate responsibility:

Be flexible like Gumby and curious like George

Working in corporate responsibility is a lifelong learning experience that rewards the flexible and curious. Corporate responsibility touches just about every issue within the company. On a single day, you may have to field questions on your company's human rights policy, the independence of the directors on your board, your water conservation measures, and the diversity of your workforce.

The breadth of this role can be both terrifying and exhilarating. The terrifying part is being asked to represent areas you know very little about. The exhilarating aspect is learning about all of these areas. People who are naturally curious, have a high tolerance for ambiguity, and are eager to take on new tasks tend to thrive in corporate responsibility roles. While you may have to spend significant time operating outside of your "comfort zone," the upside is learning about other functions within the company and building relationships with managers across the enterprise.

If being curious does not come naturally, practice by seeking out colleagues from other organizations that have a stake in your CR program. Set up one-on-one meetings to understand their scope of responsibilities, their views about your company's CR programs, and any areas of mutual interest.

Hold on to your core competency while learning new skills

Just about everyone in corporate responsibility started their careers in another field. Whether you come from a marketing background or environmental science, corporate responsibility is a big tent and there is always a way to apply your skills. The key to success is to walk the line between contributing knowledge from your core competencies, and being pigeonholed into a narrow role defined by these competencies.

Look for ways to leverage your existing skills to help the organization while simultaneously taking on other responsibilities that demand skill development. As discussed in the last chapter, my view is that people with a technical background can learn some of the "softer" skills (e.g., communications and influencing) needed for success in CR more easily than non-technical people can come up to speed on the intricacies of fields such as environmental engineering. The flipside of this opinion is that many of the people who gravitate toward technical fields may be less comfortable with the ambiguity in a CR role, and fewer of them may possess the communication and influencing skills needed for success in this profession.

While mastering new skills and behaviors in the workplace can be incredibly difficult, it can be accomplished with the appropriate time and attention. You have to be willing to take risks by working in new areas and be willing to feel vulnerable, or even fail. The keys are to have the desire to learn and grow, the humility to be less informed than others, and above all the passion for your cause.

Communicate, communicate, communicate

There is no other single skill as important as communication for success in corporate responsibility. Whether in written communications, in speaking to large groups, or in persuading a small group of internal stakeholders, communications skills are essential. The corporate responsibility professional is often in the position of communicating a fairly complex set of facts – for example, climate protection strategies – to emotionally charged, yet less technical audiences. The ability to condense complicated topics into a relevant and cogent set of messages and present them skillfully can be the differentiator for your success in corporate responsibility. Communication is such an essential element for success in this field that Chapters 8 and 9 are dedicated to the topic.

At some point, most people working in corporate responsibility will work on a corporate responsibility report. If you have ever had this experience, you know that the incoming data is from every corner of the company and authored by staff that are not necessarily creative writers. The ability to take this information and weave it into a compelling, readable report is the hallmark of a good corporate responsibility program (see Chapter 9).

If you were not born with these skills, do not despair. Communications is a trainable skill if you apply the proper focus. Personally, I dreaded speaking in public at the beginning of my career. When I recognized that this deficit was a barrier to advancement, I got some training and over time went from avoiding speeches to seeking them. Years later, I even served as an instructor for a public speaking course. Like playing the piano, you can build your communication skills through instruction and practice. Seek feedback and, combined with your own self-assessment, define the areas where you will take tangible actions to improve and practice, practice, practice! (See Chapter 8 for a full discussion of spoken communication skills.)

Lead through influence

Since the corporate responsibility professional has responsibility to communicate about numerous programs that he or she has little authority to control, it is essential to *lead through influence.* To effectively influence others, the ability to grow productive professional relationships is essential. In practice, this means that the corporate responsibility manager must build relationships with staff from departments across the corporation in order to align their actions with the strategies, goals, and metrics that make up a credible, consistent program. Like communications, not everyone is born with this skill, but it can be learned.

Leading through influence means building relationships with internal stakeholders, which can be tricky. For example, if your company's procurement vice president sees no value in auditing the supply chain for compliance to the company code of conduct, your job is to help him or her to understand why this is important and valuable.

The discussion could include everything from reducing reputational risk to your company's brand, to improving product quality, to just doing the right thing. By taking the time to understand the organization's issues and priorities, you can adjust and tailor your proposal to fit their paradigm. Most functional group leaders are not keen on "outsiders" coming in and telling them what to do. Accept that you will not successfully influence internal stakeholders in every case, but never give up. The key to success is to demonstrate *mutual value* in the relationship.

The most important thing to remember is that the relationship is more important than any short-term victory. When confronted with resistance, look for incremental steps toward your goals. Solidify the relationship with recognition of your partner's achievements in your company's corporate responsibility communications and more progress will likely follow.[24] It is essential to give your internal stakeholders *all of the credit* for incremental corporate responsibility moves. While giving credit works better for some than for others, there is no faster way to alienate your internal stakeholders than to create the perception that you are taking the credit for their actions.

Another surefire way to build relationships with internal stakeholders is to be a good listener. Ask open-ended questions that prompt the person to talk about what they do and the problems they face. Adopt a consultative approach: genuinely seek to understand their business, their issues, and, where appropriate, offer to help them with solutions. The opposite of this approach is to come in to an internal stakeholder discussion with your agenda at the forefront of the conversation. What the listener hears is "here is more work that I need you to do" and likely reacts with "who gave this person the authority to give me more work?" While this approach can sometimes yield short-term results, it is not a relationship builder and can lead to a disaffected or even hostile business partner down the road.

Remember that the corporate responsibility manager is dependent on the actions of others for his or her success. Unless the leaders of key business functions within the company support your program's strategies and goals – and take action to implement them – there will not be much substance to your corporate responsibility program.

Read the system

In Chapter 1 I used the metaphor of the epiphany and the 2x4 to describe how companies come to adopt corporate responsibility as a function. An important skill for the corporate responsibility manager is to understand the overarching paradigm of the business they are in and how it shapes corporate behavior.

The skill here is harder to define because it refers to understanding what is not being said. For example, you might have a meeting where you come away feeling fantastic about the commitment to corporate responsibility goals, but a few months later you realize that nothing has been accomplished to implement these goals. In some corporate cultures, people would rather appear to agree, when there is some unsaid reason why they do not agree. Another common example is when people sign up for one of your regular meetings on corporate responsibility, but rarely attend or appear to be checked out throughout the session.

After a few of these experiences you need to ask yourself what is really going on. While there may be no single answer, success lies in asking the question and analyzing the situation. Depending on the circumstances and the relationships, you might directly ask the person or group why they are not attending, but often you might be better served by approaching the issue through a confidant who can help you understand the situation.

Once you do understand the "system" – or the reasons behind the observed behavior – there is a range of ways to respond. Of course, all of the possible solutions (ranging from doing nothing to escalating to your management chain) depend on the situation and the relationships.

While this skill area may seem like the least defined, doing this poorly can be the most deadly to your career. If you are not able to discern the paradigm of the group or individual you are trying to influence, or if you are unable to recognize the roots of your own behavior, the outcome can be disastrous for you.

For example: a young, eager, assertive corporate responsibility manager is driving a resistant plant manager to improve the water consumption performance. Over and over, the corporate responsibility manager highlights the poor performance of the plant manager's facility and, in doing so, puts him under pressure to improve. Behind the scenes, the plant manager is also under pressure to reduce operating costs as part of a CEO challenge to boost profitability. The corporate responsibility manager ups the ante and highlights the issue in front of others to try to drive the plant manager to take action. The plant manager feels backed into a corner and he escalates the situation to executive leadership to understand where he should apply his resources. When confronted with the facts, executive leaders prioritize cost-cutting over water-use reductions.

At her year-end performance review the corporate responsibility manager is given a very strong negative message owing to negative feedback from the plant manager. Regardless of any of her other spectacular achievements, the corporate responsibility manager will likely get no raise, and no bonus. Instead she may be hanging onto her job by a thread because of the conflict she caused by pushing this issue to the brink without taking the time to understand "the system." Regardless of how well intentioned her motives were, it may take this manager years to overcome the reputation as a "bull in a china shop" due to this one incident.

The ability to understand the system takes emotional intelligence, or the ability, capacity, and skill to assess and control the emotions of oneself, of others, and of groups. While it may seem touchy-feely to some, many companies have recognized the importance of these "softer skills" and are actively training their employees to understand their own personal traits, and the signs they get from others. A common first step is a self-assessment called the Myers-Briggs Type Indicator (MBTI). The MBTI is a questionnaire designed to measure psychological preferences in how people perceive the world and make decisions. By understanding how you and your colleagues process information, you are better able to decipher how different groups and individuals operate to determine the most effective course of action. You can read more about emotional intelligence at www.eq.org.

Learn and practice 'corporate jujutsu'

A common mistake that many corporate treehuggers make is to be a bit too passionate about their cause to protect people and the planet. This can come across as overzealous and might communicate a lack of understanding or commitment to other corporate imperatives.

Working on social and environmental causes within a big company is a bit like corporate jujutsu. Jujutsu is a "Japanese martial art and a method of close combat for defeating an armed and armored opponent in which one uses no weapon, or only a short weapon."[25] While the martial arts metaphor might seem confrontational, success in jujutsu is defined as being "gentle, supple, flexible, pliable, or yielding" (*Jū*) and "manipulating the opponent's force against himself rather than confronting it with one's own force" (*Jutsu*).[26]

Working in the field of corporate responsibility, you will often find that the best pathway to achieving your results is a circuitous one. There will be numerous times when you are told "no" and given a ton of reasons why your ideas or programs won't work. As discussed in Chapter 1, a defining characteristic of corporate responsibility is working in areas where you have no authority. In this situation, you will frequently encounter resistance to your CR programs and initiatives. The essential skill here is to absorb the force of the resistance with grace (*gentle, supple, flexible, pliable, or yielding*) but stick with your values and find creative alternatives to continue to work with your business partners to achieve success (*manipulating the opponent's force against himself rather than confronting it with one's own force*).

For example, an increasingly important issue for corporate responsibility programs is working in the supply chain for your company. As outsourcing has increased, so have the expectations that your company's suppliers adhere to a code of ethical conduct. When you take these issues to the vice president in charge of procurement, you might encounter stiff resistance. Typically procurement departments are run by shrewd negotiators who optimize price, quality, and time to delivery – the procurement VP may have little time or patience for driving suppliers to comply with environmental and labor standards or, even worse, he or she might see these issues as additional costs or a source of added delays.

Rather than confront the procurement VP, one way to approach this kind of resistance is to make it as painless as possible and demonstrate that you will add value to their program. For example, if you offer to do the work to weave corporate responsibility issues into existing business practices and demonstrate that this can drive increased efficiency or reduce risk, you might be able to gain a foothold. As outlined in Chapter 7, you might offer to conduct social and environmental performance assessments for top-tier suppliers in the regularly scheduled supplier business review meetings. The central theme in practicing corporate jujutsu is to avoid head-to-head conflict with essential business partners. Instead, listen, seek to understand their issues and concerns, and then develop creative ways to work cooperatively with your business partners and weave responsibility into the relationship for mutual benefit.

Be entrepreneurial

Success in corporate responsibility often means finding hidden value. The notion that a company can simply contribute to charity, volunteer in the community, or reduce its environmental footprint and expect to be a leader in corporate responsibility is long dead. These days, successful corporate responsibility programs are integrated into the business which means that corporate treehuggers must be entrepreneurial and find the most efficient and effective ways to return value to the company. While corporate responsibility programs are often given a pass on the company's ROI test, the pressure is mounting for these programs to add more value to the business.

The leaders in integrating corporate responsibility into business are the "epiphany" companies that have made societal benefits part of their value proposition. These companies are applying their core competencies to address social needs and, in the process, turning a profit. Whether you work for an epiphany or a 2x4 company, however, the need to tie corporate responsibility to business value is an essential skill. Don't assume that, because you have a job in the CR department, your company's leaders accept the value of your role. Success depends on being able to find, assess, and prioritize initiatives that can add value to the company. This can be done from any level in the company, but requires an entrepreneurial mind-set.

For example, a successful technology firm has been criticized by the public for being lax on taking back its products at the end of their useful lives (sometimes referred to as extended product responsibility). NGOs have mounted publicity campaigns to show that these materials have toxic components and end up being disassembled in terrible conditions in Chinese sweatshops that pollute the environment. While this situation could become a PR disaster for this company (and has been for several technology companies), an entrepreneurial corporate responsibility manager could turn it into a value-producing program. Rather than simply starting a take-back program, the company response to this challenge could include strategic partnerships that increase business and brand awareness. By partnering with technology retailers or charities, the company could set up drop-off sites for used equipment to be reused or properly recycled and reward its eco-minded customers with discounts on the purchase of newer models.

Marrying corporate responsibility with the profit motive is the Holy Grail for the corporate treehugger. Like in the example above, if you can be the connector that brings the people together to develop new programs that benefit the company and the environment, you will become a superstar. The key to success in this area is to be an idea generator. Look for ways to connect the dots within a company. Look at social/environmental problems as potential business opportunities. Ask yourself, "How could we apply my company's core competencies to help solve a social/environmental problem?" Apply business acumen to these problems and you may discover the program that differentiates your company, and you, as a leader.

Doing this well requires comprehending a couple of basic truths: First, you need to understand that ideas are cheap and implementation is expensive (another way to say this is that all ideas are great ideas until you have to pay for them). Not all ideas are winners and you need to have a thick skin to pitch ideas to decision-makers who will often reject them out of hand. Look for the ideas with the best return on investment and, above all, keep innovating until you find the best approach.

Second, it is important to understand that strategic ideas are not the sole province of the executive suite. In fact, in many cases, good ideas bubble up from within the organization. Any person at any level can be the spark that ignites the next great corporate responsibility program. Often good ideas incubate in a "skunk works"[27] of co-collaborators who work under the radar to flesh out the concept before going through the normal management approvals (where most ideas go to die).

A few great reads on the topic of integrating corporate responsibility into business come from Harvard Business School Professors Michael Porter and Mark Kramer. Specifically, their papers titled "Strategy and Society: The Link between Competitive Advantage and Corporate Social Responsibility"[28] and "The Big Idea: Creating Shared Value"[29] are essential reading for the aspiring entrepreneurial corporate tree-hugger. Also, to understand the paradigm of being an entrepreneur within an existing business there is no better book than Clayton Christensen's *The Innovator's Dilemma*.[30]

Attention to detail, discipline, quality, and results

While this topic may seem like the basis of any successful program, I am continually amazed by how often these basic building blocks are ignored, especially in corporate responsibility programs. It is fun to focus on the "shiny and new" topics that are part of any corporate responsibility program. For example, it is easy to get drawn into the latest NGO "name and shame" campaign, the latest government regulation, the newest responsibility ranking list, or the flashy design of your competition's new corporate responsibility website. But, in overreacting to the hot issues of the moment, there is an opportunity cost: losing focus on the sustaining elements of your program.

An important attribute to any successful corporate responsibility program is a disciplined management system. This means that you need to understand and clearly communicate the measurable goals that your program will deliver each year and develop the business processes that will produce these results. For example, producing an annual corporate responsibility report requires that you establish public-facing goals for each program (e.g., water use reduction, workforce diversity, supply chain audits, etc.), collect the data needed from each department, and create a compelling story for the report. To succeed at this year after year, you need to be able to establish the basics: who is providing the data, who will write the story, who will review, who is providing all of the graphics and images, as well as manage the schedule and budget for the entire endeavor.

These tasks are not sexy, but doing them well is essential to running a successful corporate responsibility program. There is a common misperception about the amount of tactical work involved in managing a corporate responsibility program. Like any other corporate endeavor, the people who work in these departments often have too much to do at any given time and can be stressed. Having a disciplined program, with well-understood goals, clear roles and responsibilities, and reliable business processes will reduce this stress and produce better results. A proven success strategy is to develop a "dashboard" of key performance indicators to track the performance of your program.[31] In Chapter 4 we will explore a few fundamental techniques for keeping your program on track and measuring the right outcomes.

Again, don't assume that, because you may be new to the company or the job, someone else must have thought through all of these processes before. Corporate responsibility is often so new that the basic management systems are not yet in place. By organizing the programs, processes, and data in a disciplined way, you can add a lot of value.

Above all, passion for the cause

Whether you add value by developing detailed management systems or by running an entrepreneurial skunk works, the common element for most people in corporate responsibility jobs is passion for the cause. If you look at your profession as a cause rather than a job, you will find the energy to persevere through almost any situation. Regardless of your background or skills, the common denominator for most CR professionals is a passion to do something wonderful: to help people and the planet and to leave a legacy of a career dedicated to making the world better.

All of this may sound trite, but it is absolutely necessary. There are many hurdles to being a treehugger in the corporate world, and there will be many days when you might question this career choice. But ultimately, when you can connect your time and talent to something bigger than yourself, you can achieve deep and profound satisfaction in your career. The key to success as a corporate treehugger is to nurture the flames of your passion even when the inertia of company bureaucracy douses it with cold water.

In the Introduction, I quoted Gary Hirshberg (the CEO of Stonyfield Farm) in his speech at the 2010 Net Impact conference saying: "You should never compromise your values for work. If a company makes you check your values at the door, find somewhere else to work."

While that is a memorable and passionate statement, in reality many treehuggers do not have the luxury of being able to fully express their values at work. This does not mean that you have sold out on your values by working in a corporation. Every company is at a different point in its journey to being more responsible and ethical, and you can make a difference from any position if you are passionate, persistent, and patient. In fact, you might be able to achieve more by helping a laggard company than a responsibility leader. Regardless of where you work, hold on to your values, and use the lessons from this book to strategically and incrementally move your company in a positive direction and achieve true career satisfaction.

Ultimately the skills and attributes described above can be applied to add value to many career paths within a company. This is a good sign, because it speaks to the integration of the corporate responsibility function into business. Rather than being a career cul-de-sac, a position in corporate responsibility cultivates skills that are applicable to a broad spectrum of career paths.

3
Setting the strategy

This chapter outlines how to design a robust corporate responsibility program, including a step-by-step approach to establishing clear strategies and objectives, as well as how to predict and manage emerging issues.

If a man knows not what harbor he seeks, any wind is the right wind (Seneca).

You have a role or are thinking about a role in corporate responsibility. What do you do first? How do you prioritize all of the issues that compose the huge scope that you must cover? How do you accomplish all that needs to be done with the limited resources that are the norm for corporate responsibility departments? In this chapter we will cover methods to establish priorities, set clear goals, and distribute the workload. We will also outline techniques to identify emerging issues to ensure your programs stay on the leading edge.

Materiality

Working in corporate responsibility can be a lot like being the plate spinner at the carnival – you are constantly moving between projects and disparate topics that are important and without your care and feeding may fail. Developing a successful corporate responsibility program requires that you start with a clear strategy based on a few critical, high-priority issues. Corporate responsibility practitioners call these "material issues" and the technique to identify these issues is a "materiality analysis."

The starting point for any strategy is defining your priorities. While this may seem obvious and simple, in practice it can be complex and challenging, because the field of corporate responsibility is so broad and all of the issues seem important. The business axiom that applies here is: "When everything is a priority, then nothing is a priority."

The materiality analysis is a structured process to distinguish the most important issues around which you will build your strategy. It is a relatively straightforward process of aligning your company's priorities with the concerns of stakeholders external to your company. While there are several versions available, Figure 1 is a generic representation of the four quadrants in a standard materiality analysis. The two-by-two matrix is bound by the horizontal axis indicating increasing importance to influential stakeholders outside of your company. The vertical axis indicates increasing importance to your company's business success. Like all two-by-two matrices, the items that end up in the top right quadrant are the areas for additional focus. Does this mean that you can ignore the CR issues in the other quadrants? No. The materiality analysis is a priority-setting tool that provides you with a small subset of opportunities that constitute the primary focus of your company's corporate responsibility goals, investments, programs, and communications. This small set should be considered as potential leadership opportunities for your program – in other words, the issues where additional investment could return good value for your company *and* produce demonstrable societal benefits.

As your company and your CR programs mature and evolve, you should periodically review and revise the materiality matrix. It is wise to revisit your materiality analysis on an annual cycle or if there have been significant changes inside or outside of your company.

Figure 1 Corporate responsibility materiality matrix

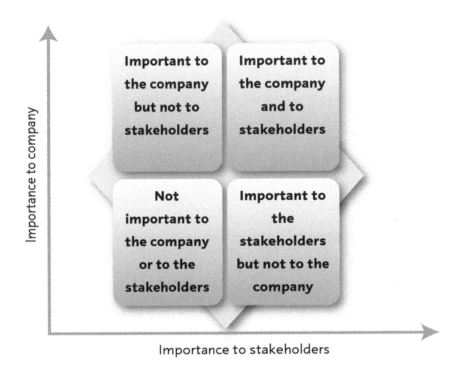

The following is a list of steps for effectively applying a materiality analysis in the development of a corporate responsibility strategy:

Step 1: List the issues

Create a list and description of the universe of issues that could fall into the realm of corporate responsibility at your company. This can be a daunting task since the field is so broad, and it is difficult to have knowledge or expertise in all of the issues. A partial list of the issues that are within the scope of a corporate responsibility program includes:

- Ethical treatment of labor in the supply chain
- Corruption and bribery in the supply chain
- Product toxicity concerns
- Product energy efficiency
- Health and safety for workers and customers
- Diversity and equal opportunity
- Product packaging, take-back, and recycling
- Carbon footprint and climate protection
- Water scarcity
- Philanthropy and volunteering
- Business ethics
- Independence and diversity of the company's governance structure
- Special risks such as nanotechnology or toxic materials
- Resource consumption
- Privacy of customer information
- Public policy and government lobbying
- Human rights
- Security practices
- Labor management relations (e.g., freedom of association and collective bargaining)
- Protecting biodiversity
- Economic impacts
- Sustainability innovations/products
- Compliance and fines

Obviously, there is a long list of issues that fall within the purview of corporate responsibility. In this initial step, gather a robust inventory of all corporate responsibility issues. A good starting point is the Global Reporting Initiative (GRI)[32] list of performance indicators. In addition to the GRI, review the criteria for awards and rankings (see Chapter 13) such as the Sustainable Asset Management's (SAM) 85+ page questionnaire used to screen companies for inclusion on the Dow Jones Sustainability Index.[33] These lists and surveys will provide you with insight into the corporate responsibility issues that external stakeholders are monitoring.

An efficient way to approach this step in the analysis is to hire a consultant that has experience with these analyses within your business sector. The consultant should be able to provide you with a comprehensive list of corporate responsibility issues that similar companies have faced, and help you with the prioritization process.

Step 2: Narrow your list

The next step is to thin down the list. For example, worries over childhood obesity might be a high-priority issue for a company like McDonald's, but are not applicable to a company like Apple. Anyone with knowledge of your company's business model can winnow the list down to the applicable issues in a couple of hours. It is often best to convene a small core team to work on this step. If you have a standing corporate responsibility committee, you could select a few people from different departments and walk through the list with them to see which issues might be eliminated before you start the prioritization process.

Step 3: Prioritize your company's issues

Figuring out the priority corporate responsibility issues from your company's perspective is the next step. The best method to tackle this task is to set up a series of interviews with the business leaders from each major department/business unit in your company. Target these interviews to the executive staff if you can – including the chief executive officer. Interviewing top executives can be challenging, but it is essential for a robust strategy process. Depending on your company's structure, there will likely be a couple of management layers with whom you will need to coordinate to set up the interviews. If interviews with the executive team are not feasible, shoot for the highest level of management possible in each relevant business unit. Perhaps the top executive might delegate someone for the interviews to represent their area of responsibility.

Start by setting up meetings well in advance, with a detailed agenda. The agenda should outline the purpose of the meeting (to discuss priorities for the corporate responsibility strategy), the process you will follow, pre-read materials, the list of questions you plan to ask, and the narrowed down list of corporate responsibility issues that will be discussed. With most of the people you will interview, this preparation will make the process go much more smoothly, but expect that some people will show up to the meeting without having read the materials. For this group, you should develop and practice an "elevator speech" – a concise explanation of the purpose of the meeting, the process, and how you plan to use the results.

Expect reactions to range from clueless to eager, and adapt your interview style appropriately. Construct your interviews around open-ended questions,[34] and you will gain a much greater understanding of the issues, plans, and concerns from each one of the people you interview. If you can, it is best to have someone accompany you to take notes so that you can focus on the interview. In many cases, bringing in a consultant to conduct the interviews and/or take the notes can be extremely useful because it gives you a buffer that allows you to listen and learn from the interviews. If you use a consultant, you should still participate in each of the interviews because you will not only gain first-hand knowledge of company priorities and perspectives, you will also build relationships with business leaders who are essential to the success of your corporate responsibility program.

Another method, although less advisable, for eliciting corporate responsibility priorities is a survey. Surveys can be useful, but they tend to be overused in the corporate world. While you might get more responses this way, they may not be well-considered responses. The trend in most companies is to send too many surveys to employees on a myriad of issues. Many employees feel over-surveyed and don't think too much about the answers they give. Further, the survey taker is usually very busy, trying to clear away yet another task in their stuffed inbox, and is likely not familiar with the issues. Thus, most survey results are compilations of a series of snap judgments, and may not be representative of actual opinions.

Nonetheless, surveys can be useful tools if used judiciously. Avoid over-surveying, over-analyzing the results, or solely relying on the survey responses as the absolute truth. Surveys results must be "sanity checked" by the person/group responsible for the strategy. In other words, never feel constrained by the results of a survey – not every data point is accurate and you should corroborate the information and apply judgment when making decisions based on the results.

Step 4: Prioritize your stakeholders' issues

The outcome of Step 3 should be a clear understanding of the priority of the applicable corporate responsibility issues from the perspective of your company's leaders. The next step is to gauge the expectations of external stakeholders. This can be the trickiest stage of the process, as each stakeholder group may have a different agenda. Again, a consultant with experience in this process for your industry can be helpful. Seek out consultants that have experience working with your industry and have frequent interactions with NGO activists, socially responsible investors, and other stakeholders that track corporate responsibility performance in your sector, or, better yet, your company.[35] This can be a wise investment since these consultants should be able to provide you with a one-stop shop for solid and credible ranking of the issues that are likely at the top of the priority list for your stakeholders.

Another approach is to establish a stakeholder advisory panel. If you have the time, inviting external stakeholders to provide input on the selection of priority issues can be extremely valuable. Their perspectives and insights can be richer and more impactful if you are able to engage directly. Stakeholder engagement, however, can be a lengthy and delicate process, as explained in Chapter 10.[36]

Short of a formal stakeholder panel, you can informally interview a few influential stakeholders within your network. Again, outreach and discussion will help to build relationships. The preparation is very similar to the internal interviews, but be sure to establish some ground rules up front. For example, you should set out whether the stakeholder comments are anonymous or can be attributed; develop a common understanding of how their opinions will be used as input to the strategic process; and determine how and when you will provide feedback to the stakeholders on the outcome of the process.[37]

Step 5: Pulling it all together

Perhaps the most important element in the development of a robust materiality analysis is a group meeting of your internal stakeholders. Owing to the breadth of the issues covered by corporate responsibility, many companies set up a council or advisory committee to guide the program. If your company has an existing council, this is the team that should be gathered together to develop the final list of priorities. If your company does not already have a corporate responsibility council, the development of a new strategy (or a strategy refresh) is a great opportunity to set one up. Similar to the executive interviews, the council membership should be made up of senior-level people from the key business units. In fact, during the interview process you could ask the executives to assign a delegate to the council. The ideal candidate is someone who is in a decision-making role within his or her business unit, has a passion for the topic of corporate responsibility, and can devote the time needed. An effective CR council is an important mechanism to help sort through and prioritize complex and sometimes ambiguous issues.[38] Like a jury, people will bring in their unique perspectives, and the group conversation will spur creativity and thought that would not have happened otherwise.

To have successful meetings on establishing corporate responsibility priorities, it is essential to be well prepared. When setting up the sessions, it is best to set aside at least four hours in one of the more comfortable meeting rooms in your offices, or consider setting up an off-site location. Although off-site meetings can be more difficult and costly to set up, an off-site location helps participants to mentally untether from their day jobs, resulting in better focus and engagement. Again, send out pre-read materials with clear expectations, an agenda, and relevant background documents and be prepared for some of the participants to have skipped their homework.

I have found it best to start these meetings with a review of the objectives, the plan for the day, and – as an icebreaker[39] – go around the room and ask each person to state their expectations for the meeting (capture these on a flip chart to refer back to at the end of the meeting). After this, the opening presentation should cover the results of the internal and external interviews and set out the initial findings of the materiality analysis. Keep it short so that the audience does not slide into listening mode (also known as "thinking about something else mode") and to ensure that the audience knows that you are not giving them the answer, but rather a starting point for the discussion to follow.

In my experience, there is a huge value in hiring or assigning a facilitator for the meeting. If you have to manage the whole session, you will not be able to participate and, worse yet, the topic can appear to be a niche or even a personal agenda. Have the facilitator speak first and establish themselves as the emcee by setting out the ground rules for the session. Assign speaking roles to several people, perhaps bring in some outside speakers if the time and budget allows, and – most importantly – save plenty of time for group discussions.

Group discussions are the heart of the meeting. This is where you will get the perspectives and discussion that will establish the priorities for your program. These can be done as a facilitated dialogue with the whole group or, if you have a larger group, split the attendees into smaller breakout sessions. In either case, set out clear instructions, such as "evaluate the results of the interviews and define the top three highest priorities for the corporate responsibility program." Draw up your list of the questions for the group discussions beforehand and ask them if they understand the expectations before you begin. If participants are not jumping in with ideas, ask someone directly or poll the group by asking each person in the room to respond to the prompt. Make sure that someone is capturing the input of the discussions on a flip chart.[40]

Follow up the group discussions with a summary of the key points. It is during the summary that the group's thoughts start to crystallize into a common position. Look for and encourage the minority report – the views of the iconoclast in the room who sees things a little differently. Seek out the quiet ones who have listened a lot but not offered too much verbally. You might find that "still waters run deep" and the quieter participants – those who have spent more time listening than talking – will come up with the observation that can completely change the direction of the meeting.

The tough part is bringing it all together at the end. This is situational, and takes skilled facilitation – someone who can weave the common threads across diverse perspectives and draw out the themes that will have everyone silently nodding in agreement. Depending on how the session has progressed, the closing can range from "we have a couple of important ideas, but will need more time to work through the issues," to "here are the priorities we have agreed to today, let's talk about next steps."

In my experience, I like to take notes on the common themes that emerge throughout the day. For example, at AMD, the corporate responsibility council had recognized business ethics as a material issue. Each time this issue came up though, it was clear that the group thought we already had a well-managed and -resourced program. This observation led the group to a paradigm that helped us find the signal in the noise: we added another criterion to our priority setting – issues that are "important but well covered." This quickly surfaced a series of other issues that were "important and need more attention." Once we agreed on this additional criterion we were able to establish a small and workable list of issues for additional focus.

Don't rush the materiality assessment. It may take a series of meetings and follow-up discussions. As discussed above, if you can, you should engage external stakeholders in the materiality process. You should also review the draft results with the executives you previously interviewed before finalizing. The people in positions of authority over the business groups responsible for the material issues that you identify must agree with your assessment and accept their responsibility to address these issues.

For example, your process results in a recommendation that supply chain social responsibility is a high-priority issue. This conclusion is based on well-documented activist allegations of sweatshop conditions in your industry's supply chain, coupled with a recent shift by your company to more outsourcing. Based on the data, prioritizing this issue makes sense, but it could be meaningless unless the head of your company's procurement team buys in to the conclusion and will help drive the strategy.

One last word on the materiality meeting – or all long meetings for that matter: Serve food. The old axiom that we learn the most important life lessons in kindergarten is applicable to long meetings. When people are hungry, tired, or both, they will be cranky and will disrupt your meeting. Not only will I serve food at long meetings, but I also like to schedule plenty of breaks and, if possible, schedule these meetings on Fridays – for some reason people are just a lot happier on Fridays . . .

Setting the strategy

Once you have defined the list of priorities for your company's corporate responsibility program, it is time to develop strategies for each. There are many books on developing business strategies, but I have found that effective strategy can be distilled into three main components:

1. **Vision.** Set out a clear vision of where you want to be in the future (this is also referred to as the future state). The vision can be "aspirational," meaning, "we have no idea how we will ever achieve this," or it can be literal, meaning, "we are fairly certain that we will get there." Avoid generic vision statements like "be the world-class corporate responsibility program," or

lengthy vision statements that try to fit in everyone's pet priority. The best vision statements are short, memorable, and actionable[41]

2. **Objectives.** Break down the steps to achieving the vision (future state) into bite-sized, achievable programs. For example, if the vision is: "Be, and be perceived as, the most responsible brand in our industry," a supporting objective might be: "Publish a GRI-certified Grade 'A' report prior to the stockholders meeting." Each objective should be assigned to an owner who has the responsibly, authority, *and* resources needed to achieve success. Each objective should have a deadline and be measurable with data (see KPIs below) that are reviewed on a regular basis

3. **Key performance indicators (KPIs).** These are the critical few "key" measures of success for each of the objectives in your strategy.[42] For example, if your objective is that all tier-one suppliers should conform to your company's code of conduct (i.e., those suppliers that work directly with your company as opposed to those who work through an intermediary), the KPI could be the number of critical non-conformance findings from audits of tier-one suppliers. It is important to understand the baseline of the KPI data, how it will be reported, and the process for review of progress toward the goal. Qualitative objectives such as "the annual corporate responsibility report was well received by influential stakeholders" can be more difficult to measure. For more subjective goals, you can establish a process of polling certain people or groups for feedback. In the next chapter, we will cover operations reviews and other forums to oversee and manage progress against goals.

Keep it simple

In strategy setting, I have found that the "keep it simple stupid" (KISS) rule is the best advice. People in corporate life view strategy setting as a high-value activity, so expect your company to have a defined process for establishing strategy that is the "secret sauce" in their success. While it is best to adapt to whatever paradigm your company uses for developing strategy, never lose sight of the fact that unless a strategy is understood (i.e., is actionable and memorable) and implemented by the stakeholders, it is meaningless. To keep your strategies simple, it is useful to enforce the discipline of articulating all of the elements of the strategy on a single page. The generic template in Figure 2 can be adapted and used for setting the strategies around corporate responsibility or many other issues.

Ideally, each strategy is assigned to an owner who takes responsibility for outlining the objectives (tasks) and owners, ensuring that the resources are in place and overseeing the progress against the goals. Since it is often difficult to get people to agree to take on additional work, this part can be especially challenging. A tactic that has worked for me in the past is to ghostwrite the strategy for the assigned owner, but have them present it to the appropriate review body. This makes sure that: (a) it gets done; (b) it covers the topics that you need; and (c) the owner understands it enough to present to others and gets the credit for the approach. Of course, it is preferable to have the owners develop their own strategies, but sometimes by putting in a little sweat equity, you can get the ball rolling.

Getting your strategies down on paper is all-important, but it is just the first step. In the next chapter we will cover proven tactics for implementing strategies, staying focused, and driving continuous improvement.

Figure 2 One-page strategy format

Topic:	Definition:	Draft: [Date]			
Owner:					
Vision: What is winning – define the "future state" here	Objectives: How we will win – list actions here	2011–2013 Plans			
		Owner	Goal	KPIs	Deadline
	1.				
	2.				
	3.				
	4.				
	5.				

Benchmarking: The lazy person's strategy

The human species has a fundamental "herd mentality." We instinctually feel safer when we are doing what everyone else is doing. We feel vulnerable when we stand outside of the crowd. Perhaps this is why one of the most popular methods for approaching corporate responsibility strategy (and all strategy) is benchmarking.

Benchmarking is studying the actions of others in your field and comparing them with your own strategies. The typical ways to approach benchmarking are to: (a) identify competitors; and (b) identify the leaders in the field both within your industry (competitors) and across all sectors; then (c) identify their actions; and (d) compare their actions against your plans to identify gaps. Benchmarking is a big business for consulting firms because it is always interesting to know what others are doing on similar issues, and especially interesting to find out about leading firms.

While this can be extremely useful information for identifying gaps in your program, it should not be the only input for establishing strategy. By definition, benchmarking is backward-looking. It typically identifies what others have already done long after they have done it. For example, by the time a company is ready to disclose its strategies and results to the outside world, it has been through the planning and implementation process and is likely seeing results. Benchmarking does not reveal the emerging issues that may impact your company and your strategy; instead it shows you how others prioritized and managed issues. While benchmarking can reveal best practice and the gaps in your programs, relying on benchmarking alone can lead you to following the actions of others.

Effective strategy setting means that you have to be part fortune-teller, part iconoclast, and a first-rate risk-taker. Look around corners to identify the approaches that have not yet been tried and tested. Use benchmarking to identify what others are *not* doing and to generate new ideas. Find emerging threats and opportunities, and then take that leap of faith to craft a unique niche that will differentiate your company. There is a tendency in strategy development to wait for complete information and absolute certainty before taking action.

Think different: The iPhone example

When I worked at Apple, CEO Steve Jobs announced the iPhone. Smartphones were prevalent at the time of this announcement (July 2007) and the mobile phone business model was well established with a few dominant players. If Apple had relied on benchmarking and emulating the actions of leaders in the field, there would be no iPhone today.

The company motto of "think different" says it all. Mr. Jobs announced the iPhone by saying, "I don't know about you, but I hated my phone." He was so sure that the iPhone would change the smartphone industry forever that he said, "You will remember where you were on this day."[43]

The point here is that Steve Jobs looked at a ubiquitous technology and said, "Hey, this sucks. I can make it much better." In doing so, Apple redefined the smartphone industry and, as a result, quickly took a leadership position in a highly competitive market. While this is not a corporate responsibility example, it demonstrates that true innovation is solving problems creatively, not simply following what others have done.

Leadership entails risk-taking, which, by definition, means that you could fail. Colin Powell (former U.S. Secretary of State and Chairman of the Joint Chiefs of Staff) provided some outstanding leadership tips, many of which are applicable to the business world. The Powell tip that applies here is:

> Part I: Use the formula P = 40 to 70, in which P stands for the probability of success and the numbers indicate the percentage of information acquired.

> Part II: Once the information is in the 40 to 70 range, go with your gut.[44]

Depending on the culture of your company, risk-taking can range from being encouraged to being punished. Since the definition of risk-taking involves the possibility of failure, make sure you understand the culture of your company and, to the extent you can, line up supporters in leadership roles to hedge your bets.

The crystal ball: Identifying emerging issues

Good corporate responsibility programs are excellent at identifying and prioritizing emerging issues. As discussed above, if your program's main source of input is tracking your competitor's actions, you will be a follower rather than a leader. But how do you uncover the emerging issues that are likely to become the next big thing and develop effective management strategies? The answer lies in developing an excellent emerging issue identification and prioritization process.

No one has a crystal ball, and all of the "futurists" are just guessing. So, if no one knows the future, what are the best ways to get out in front of issues rather than follow? There are many ways to approach this problem and none guarantees success, but I will outline some techniques that have worked in my experience.

Step 1: Issues scan

Start by defining the scope of issues you are covering in your strategy. For example, your scan could be narrowly focused on product-related environmental regulations or more broadly defined to cover the entire slate of corporate responsibility issues applicable to your company.

The issues scan is essentially a literature search of the trends and discussions within your scope. The phrase "literature search" is kind of a throwback in the Internet age – emerging issues scans involve Web searches with multiple search strings, word clouds, heat maps, social media, blogs, discussion boards, etc. – all of which can be done with internal staff, interns, or consultant support.

You may want to augment your search technique by conducting a series of interviews with people outside of your company who have a broad perspective and deep understanding of the issues at hand. For example, if the topic is environmental sustainability, talk to a green business writer from a leading media outlet, an influential NGO, or a social investment analyst. If you do interview people, design an interview guide with a few key questions that you ask consistently so that you can compare the responses. The goal is to gather information to determine which issues are being discussed by the plurality of influencers but are still nascent (which means that the issue has not yet become a government policy, widely reported in the press, or the subject of a major activist campaign). This is not an exact science and, as discussed below, the findings should be vetted in a group process.

A recent example is the emergence of water scarcity as a major sustainability issue. A few years ago, only a few influential groups were discussing this issue, while most corporate environmental programs focused on climate change. Today, water scarcity has blossomed into the "new carbon," as an issue at the top of the environmental priority list. Another example from a few years ago was the switch from focusing on the local environmental impacts of factory emissions to the impacts throughout the lifecycle of a product.

The challenge in these "crystal ball" processes is that every issue can look like a big threat or an opportunity. Of course, not every simmering issue will boil over into a real threat or opportunity that is relevant for your program. To separate the important from the merely interesting, the issues scan is just the starting point.

Step 2: Sanity check

Once you have processed the issues scan into a few trends and key issues, the next step is to convene a group process to "sanity check" your results. While this step is not absolutely necessary, it is helpful to get other opinions into the mix and can result in adding, deleting, or editing the topics that were identified in your scan.

Step 3: Gap analysis

As part of your group session, or in advance, you should compare the trends identified in the scan against your current strategies and plans to identify any gaps. For example, if your business is making semiconductors and your scan identifies emerging concerns about the health implications of nanotechnologies, the question to examine is how your business would fare if these concerns suddenly got traction in the public consciousness. Could these concerns impact semiconductor operations? If so, does your company have a plan to cope with these issues?

Step 4: Prioritize the gaps

Once the gaps have been identified, the best way I have found to prioritize them for action is to analyze the *magnitude and likelihood* of the future risks or opportunities. A good way to look at this is to ask the following question: "If we did nothing more on this issue, what is the likelihood we will be impacted (positively or negatively) by the expected trend?" Then ask: "What is the potential magnitude of that impact?" By polling a group of people with knowledge of the topic on these questions, you can get a qualitative prediction (high, medium, low) of the likelihood and magnitude of potential impacts. The product of these predictions (likelihood multiplied by magnitude) yields a fair gauge of where you should consider future investments in your program.

Scenario planning

Scenario planning is a process that elegantly ties together the emerging issue scans and strategy setting. Future scenarios are created based on your issues scan and the participants role-play as if the scenario was reality. This can be an effective and fun method to shake up participants' perspectives and paradigms. Like all group processes, it takes strong facilitation to keep the discussion on track and to extract the important lessons from the sessions.[46]

Predicting emerging issues

Intel went through an emerging issue process where the environmental team looked at the issue of climate change and saw a steeply increasing curve of public/regulatory interest (this was before the Kyoto Protocol).[45] When Intel took stock of its vulnerabilities (gaps) on this issue, the team recognized that semiconductor manufacturing was one of the few industries using materials known as perfluorocarbons, or PFCs. PFCs are a class of chemicals that are very potent greenhouse gasses – in some cases more than 10,000 times the potency of carbon dioxide.

Even though there were no regulations at the time and Intel had received no public complaints about this material, future trends pointed to emerging public outrage on the climate change issue and a high likelihood that regulation would follow. The impact of a regulation to ban or severely restrict these compounds would have jeopardized multi-billion-dollar investments in manufacturing around the world.

With a high likelihood of restrictions and public outrage, a huge potential magnitude of impact (the inability to manufacture products) combined with a lack of any plan to reduce or phase out the use of PFCs, this issue quickly became a top priority. As a result, Intel assigned an internal team to work on engineering solutions and an external team to work on climate policy. Notably, these efforts resulted in the World Semiconductor Council (WSC) developing the world's first voluntary global phase-out of greenhouse gasses. Ultimately, PFCs were regulated in several jurisdictions but, because Intel had planned ahead, it experienced no manufacturing delays and was able to collaborate with the entire industry to reduce emissions of these potent global warming gasses.

Trend spotting: Radar and sonar

While there is no single method for staying abreast of emerging trends, it is an essential capability for a corporate responsibility manager. There are two fundamental components to staying on top of emerging trends: looking outward (radar) and looking inward (sonar).

Radar

To stay on top of emerging trends and conversations on a real-time basis, subscribe to relevant blogs and e-newsletters, and follow influencers and leaders using social media tools like Twitter, Facebook, and LinkedIn. Some of the current leading corporate responsibility e-newsletters are: GreenBiz.com; Environmental Leader; 3BL; TriplePundit; Treehugger; Sustainable Brands; CSRwire; Grist; and Fast Company.

Establish a work-only Twitter handle and follow corporate responsibility influencers.[47] Participate in significant corporate responsibility events such as the Business for Social Responsibility conference; Ceres, SRI in the Rockies; the Social Investment Forum; Net Impact; Sustainable Brands; and Fortune Brainstorm Green. The goal is to stay current and understand trends.

It is also important to apply judgment and filter the issues that show up on your radar screen. Some issues look critical but may be baseless. Raising the red flag (i.e., notifying people in your chain of command) on issues that are not well founded could result in a "cry-wolf" reputation and will not be helpful to your credibility. Alternatively, waiting for an issue to become a known threat or opportunity could result in your program being too reactive and late to act. Balancing these two scenarios to identify the critical issues and take appropriate and timely action is more art than science, but it is essential and can be managed with the techniques outlined in this chapter.

Sonar

Just as important as the radar function is the sonar function, which involves tracking your company's trajectory. What you should look for are the changes that you can reliably predict and that could have a material impact on your CR program. If your company has a strategy office, arrange a time to speak with the staff and ask them about industry trends and the likely directions for your company.

For example, if your company may enter the wireless communications business, you should review the data and trends on the health effects of the wireless spectrum. You might also look at opportunities such as the use of wireless technology for smart grid applications that could reduce energy demand.

Another example of predicting company changes comes from my experience at Intel. With Moore's Law driving continuous improvements in semiconductor technology, the manufacturing processes needed to make these increasingly sophisticated devices are always on the leading edge of chemistry and physics. New and potentially harmful chemicals coming into the factories and labs meant that the environment, health, and safety team needed to anticipate and mitigate potential risks on a variety of increasingly exotic materials.

Establishing a clear strategy on the right issues with appropriate staffing, resources, and oversight is the foundation of all good CR programs. By using the techniques in this chapter, you should be able forecast and prioritize key issues and design effective strategies that will achieve the vision for your CR program.

4
Running a data-driven program

This chapter covers how to run a program based on clear objectives and data. It also covers techniques for running effective management committees, driving continuous improvement, communication, and recognition to motivate your team and internal stakeholders.

What gets measured gets managed (Peter Drucker).

In many companies that you may work for, the strategy for the corporate responsibility program is already well established. In these cases, your job will likely be more focused on managing the day-to-day operations of the program than on developing new strategies.

A lot of my training in the program management discipline stems from my career at Intel. Each new employee starts with training on the Intel values: customer orientation; discipline; quality; risk-taking; great place to work; and results orientation. After working there a couple of months, Intel employees usually re-order these values as: Results, Results, Results, and all those others.

The laser focus on getting things done at Intel drives the culture and has helped that company rise to dominance in its industry. While this culture can take its toll on the employees, there are some valuable lessons to be gained that can be applied in corporate responsibility programs, or just about any program.

Establish meaningful goals

Peter Drucker's quote, "What gets measured gets managed," is one of the overused phrases in the business world because it is fundamentally true. Establishing clear goals and regularly reviewing data that indicate progress toward those goals is fundamental to most business functions.

All corporations run on data. In public companies, the CEO has to explain the financial results and forecast every 90 days. Every department has goals and targets that are measured and reviewed. Each employee has a series of performance targets that are measured every six months. All of these measures are tied to significant consequences. Quarterly results can sink or surge the stock price (and the CEO), departmental goals determine the success or failure of an entire group, and individual goals can be the difference between getting a promotion and being laid off. Obviously, data matters in business, but using these tools effectively means setting the right goals and measuring the right things.

In a corporate responsibility program, establishing the right goals and measures is more of a challenge than for other disciplines. For example, the quality department measures the defects per quantity of products made. The procurement department measures the cost savings they can extract from suppliers. The safety group measures the number of injuries and illnesses in the workforce. The breadth of the issues covered by the corporate responsibility department can make the measuring process more difficult. Should your program track environmental performance, diversity in your workforce, supply chain audit results, philanthropy, some of these, or all of the above?

There is no generic answer to this question. The reason that this chapter follows the chapter on strategy is that *your measurement system must be tied to your strategic goals.* The broad scope of corporate responsibility requires that you prioritize the issues that are important for your business as defined by the materiality analysis covered in Chapter 3. Once you have defined your priorities, the decisions on the goals and which indicators to measure become much clearer.

The Coca-Cola Company is a great example. The front page of its sustainability website[48] lists seven issues that the company has determined are the most critical: beverage benefits; active healthy living; community; energy and climate; sustainable packaging; water stewardship; and workplace. As shown in Table 1, each of these issues is accompanied by specific, measurable goals, most of which include deadlines. The website also provides detailed data for stakeholders to track the company's performance against these goals.

Table 1 Coca-Cola Company's sustainability goals

Focus area	Goals
Beverage benefits	Invest more than $50 million in research by 2015
	Continue developing products fortified with additional nutrients to meet global consumer needs
	Innovate with natural sweeteners, which have the potential to lower calories per serving
	Strive to have low- and no-calorie options and/or smaller portion sizes available in communities where obesity is a significant problem
	List the calories/kilocalories/kilojoules per serving for our beverage products on the front of nearly all of our packaging worldwide by the end of 2011

⇒

Focus area	Goals
Active healthy living	Support the Healthy Weight Commitment Foundation in reducing the total annual calories consumed in the U.S. by 1.5 trillion by 2015
	Not directly market our beverages to children younger than the age of 12
	Support at least one physical activity program in every country in which we operate by the end of 2015
Community	Form 1,300 to 2,000 new micro distribution centers (MDCs) in Africa by the end of 2010
	Empower 5 million businesswomen in our global business system
	Give back at least 1% of our operating income annually to help develop and sustain communities around the world
	Improve the quality of life in communities where we operate by supporting key initiatives and responding to community needs through financial contributions, in-kind donations, and volunteer service
Energy efficiency and climate protection	Improve the energy efficiency of our cooling equipment by 40% by the end of 2010
	Grow our business but not our system-wide carbon emissions from our manufacturing operations through 2015, compared with a 2004 baseline
	Reduce our absolute emissions from our manufacturing operations in Annex 1 (developed) countries by 5% by 2015, compared with a 2004 baseline
	Install 100,000 hydrofluorocarbon (HFC)-free coolers in the marketplace by the end of 2010
	Phase out the use of HFCs in all new cold-drink equipment by the end of 2015

Focus area	Goals
Sustainable packaging	Improve packaging material efficiency per liter of product sold by 7%, compared with a 2008 baseline by 2015
	Recover 50% of the equivalent bottles and cans used annually by 2015
	Source 25% of our polyethylene terephthalate (PET) plastic from recycled or renewable material by 2015
Water stewardship	Assess the vulnerabilities of the quality and quantity of water sources for each of our bottling plants and implement a source water protection plan by 2013
	Improve our water efficiency by 20%, compared with a 2004 baseline by 2012
	Return to the environment, at a level that supports aquatic life, the water we use in Coca-Cola system operations through comprehensive wastewater treatment by the end of 2010
	Replenish to nature and communities an amount of water equivalent to what is used in our finished beverages by 2020
Workplace	Achieve a 98% performance level for Company-owned and -managed facilities upholding the standards set in our Workplace Rights Policy by 2015

Starbucks is another fabulous example of clear goals and measurement. The company's Global Responsibility Report: Goals and Progress 2010,[49] bundles its goals and measures under five areas: coffee purchasing and farmer support; community involvement; recycling and reusable cups; energy and water conservation; and green buildings. Each of these areas has a set of elements displayed on the webpage. When you move your mouse over any of these elements the company's goals light up. For example, under coffee purchasing and farmer support, the goal for coffee purchasing is "100% ethically sourced by 2015" and you can click on the goal to see Starbucks' current progress toward this goal.

There are a few important take-home messages from these two examples:

- Each company has set goals in the areas that are of most importance to their business and to their stakeholders – the material issues

- The goals are Specific, Measurable, Attainable, Realistic, and Time-bound (S.M.A.R.T. goals)

- They measure only one or two key performance indicators (KPIs) to determine performance against their goals – they keep it simple

- They publicly report the progress against their goals even when they have missed the goal. For example, Starbucks' electricity use goal for company-owned stores was 25% reduction by 2010, but they report that they only achieved a 1.6% decrease

Again, corporate responsibility programs cover a huge scope of issues, almost none of which is under the control of the corporate responsibility manager. This dynamic makes goal-setting far more challenging than just picking the right issues and setting targets. To assemble a robust set of goals and measures, the corporate responsibility manager must work with the leaders of the essential functions within his or her company. As discussed in Chapter 2, you will have to "lead through influence" to align your colleagues on a common set of goals and the business process to measure progress.

Start by taking inventory of the goals and measures that have already been established and relate to your program's material issues. Leverage your relationship with the program owners to get access to the processes that create the data and gain permission to utilize it in the corporate responsibility program and/or the public report. To the extent possible, integrate with the program owners so that you can influence the review of the data and the revision of the goals.

Issues that are material to your corporate responsibility program, but are not adequately measured, are more difficult. You will need to work with the program owners on the establishment of new goals and measures that may not fit within their current priorities or budget. This alignment should occur at a senior level in the strategy-setting process. In reality, the corporate responsibility strategy alone may not be enough to change behavior within crucial business functions. The best advice in these situations is the "corporate Jujutsu" technique discussed in Chapter 2, which in essence means patience and perseverance.

Measure the right things

In the last chapter we discussed the selection of key performance indicators, or KPIs. The important messages are outlined in the abbreviation KPI:

- **Key**: measuring the "critical few" items

- **Performance**: measuring the items that tell you about performance toward your goal

- **Indicators**: measuring items that indicate the performance of the system (as opposed to all system metrics)

When choosing KPIs, it is important to be very selective. This is very important because the decision to measure each KPI sets a chain of events in motion that is costly and time-consuming: the data must be collected from multiple sites around the world, quality-controlled, aggregated, analyzed, and (the most costly of all) managed. It is critical to resist the temptation to track everything and anything that can be measured.

Look for the KPIs that are true measures of the system that you seek to manage. Most indicators will be lagging indicators – in other words, measuring things that have already happened. Try to mix in a few leading indicators, or those that forecast where the system is heading.

Choosing KPIs at Intel

With 17 manufacturing facilities spread across the planet, the scope of the Intel environmental program was massive. Soon after accepting the corporate environmental manager position, I met with senior management, staff, and internal stakeholders (e.g., factory managers) to reach agreement on the KPIs that we would use to measure the company's environmental programs. The KPIs were focused on air, water, and waste emissions, resource consumption, compliance, and incidents. We established goals for each program area and divided the KPIs into leading and lagging categories. Each quarter, my team collected the data from staff in the facilities around the globe, compiled and analyzed it, then assembled a few compelling graphs. My job was to review the information and distill the critical messages. For example: Are compliance systems working? Are the facilities on track to achieving their goals? Is the aggregate of all the data on track with the corporate goals? Are there any disturbing trends? Do our forecasts suggest that we need to do anything differently?

It took a few cycles for us to establish a smooth business process, but once it was in place, these operations reviews were *the single most important management process for achieving environmental leadership.* This process provided a real-time gauge and forecast of the critical issues affecting the environmental performance of the company. Regular reviews with the right decision-makers enabled swift decisions and kept us on track to achieve the company's goals.

For example, a lagging indicator could be the injury rate in your company's workforce over the last calendar year. The corresponding leading indicator could be the number of "first aid" cases, which can be a predictor of more serious worker injuries. You should "stress test" your KPIs by asking, "What would change if we did not measure this?" Or, "When was the last time we took action based on this KPI?" If the answer is little or nothing, then drop the KPI. The goal is to track the most meaningful indicators, not those that are easy to measure.

Operations reviews

While the examples from Coca-Cola and Starbucks above focused on publicly reported goals and measures, most companies use a larger set of goals and KPIs for internal program management. Well-managed programs operate from a "dashboard" of KPIs and grade these against the goals for the program – typically as "green" (on track or ahead of schedule), "yellow" (data is in the right direction but there are problems that need attention), or "red" (the data is off-track and problems exist that may result in missing the goal).[50]

Once you have established your goals and KPIs, you will need a system to monitor and manage the results. Typically, a senior management committee (covered in the next section) reviews the data in an operations review or "ops review." Ops reviews are typically held on a quarterly basis, but the frequency can be adapted to the program's needs and culture of the company.

Well-run ops reviews are the lifeblood of a data-driven program. This is the forum where your program will be evaluated and decisions will be made that can change your priorities. You will need to spend a significant portion of your time planning and preparing for these reviews.

Step 1: Gather the data

Identify the data providers for each KPI you will track and establish a schedule and process by which you will receive information from them. If you are collecting data from a system that is widely distributed, such as a network of facilities around the world, you will need to factor in the time it takes to collect this information and run through a quality control process.

It is wise to establish a backup plan as well as a rapid inquiry process. The backup plan is essential when the person responsible for providing the information is not available but the data is still needed – there must be another person with the knowledge and capabilities to step in when needed. Rapid response is needed when the inevitable questions or quality control issues crop up. For example, you notice that the amount of water used by a particular factory has shot up since the last quarter. You are scheduled to present the data next week, but don't have the story behind why the data changed or the possible corrective actions.

A useful tool for managing this process is to document the relationship with other departments with a "service-level agreement" or SLA. The basics components of a SLA are:

- **What** will be delivered?

- **Who** is responsible for delivery?

- **When** will it be delivered?

- **Contingency plans** for people and/or problems with the data

Step 2: Prepare and analyze

The schedule ends with the operations review, but you will need to plan a series of milestones prior to this meeting to analyze the data, understand the key issues, and develop your presentation. Every ops review I ever conducted encountered some issue or question that took time to sort out. It is essential for your sanity and credibility to build in enough time to work though issues that inevitably pop up. For example, if water use per product unit increased by 20% during the quarter – moving the KPI from green to yellow – you will need to understand why that happened and outline a response plan *before* your ops review. In some company cultures, the management may be more comfortable letting the data speak for itself in a real-time discussion on the reasons, ramifications, and actions during the ops review. In my experience, it is best to fully understand the root causes of any anomalies and prepare all of the stakeholders beforehand. Without this preparation, you risk embarrassing allies who are responsible for the data, but may not be briefed on the situation.

To ensure you have time to work though all of the issues that can crop up, build a schedule that allows adequate time for data acquisition, quality control, and – most importantly – analysis. It is in the analysis phase that data turns into information. Using the water consumption example above, you might be able to explain the increase by looking into process or production changes that may have occurred over the last quarter. It is essential that each finding that is presented to senior management has a story – a take-home message – that explains what is going on within the program. This takes time to figure out, which must be built into your schedule.[51]

I recall being asked very obtuse and unexpected questions about the data we presented in the environmental ops meetings at Intel. The same discipline was applied at Apple when (then COO) Tim Cook grilled me about minutiae each time I presented the results of the supplier responsibility program. Sometimes it seemed like these questions had more to do with testing my grasp of the program than the actual results.

There is no substitute for knowing your stuff. While you may only communicate a few high-level messages to senior management, you should absolutely be prepared to dive into the details with a masterful understanding of all issues within your scope at a moment's notice. Doing your homework not only helps you answer the unexpected questions and demonstrate that you have a grip on your programs, but it also helps you rein in any unqualified conclusions based on the data. For example, some of the environmental emissions data at Intel was based on formulas we applied to chemical purchasing records. This meant that there was some variability and uncertainty with the actual emissions. If we saw unexpected results, we would dig into the data and often found reasons like end-of-quarter bulk purchases that explained the anomaly.

At Apple, a colleague advised me about ops meetings with Tim Cook this way: "I approach these meetings like I am studying for a final in college." Mr. Cook is known for having an incredible memory and grasp of details.[52] He has an uncanny ability to find the one question that you did not anticipate. You should expect that a senior manager at your company will test your knowledge of the details behind your presentation and you should be prepared to respond. You should also anticipate being stumped by at least one question (I have encountered senior managers who will continue to ask increasingly detailed questions until you run out of answers). In these cases, the worst thing you can do is to waffle or attempt to BS the senior manager by making something up. Candidly admit you don't know and offer to follow up with the answer.

Preparing well for operations reviews requires that you understand the source and meaning of every data point. It is a lot like a test in college – you study! When you think about it, senior management has a much broader scope of issues to manage and, by asking you probing questions, they establish confidence that you understand and are managing your area.

Step 3: Tell the story

Your analysis of the data should lead to conclusions that have consequences. If assembling a set of quality performance indicators is science, then figuring out what the data is telling you is art. For example, the safety incidents at your company's owned facilities are low and tracking lower every quarter for the last eight quarters – does this mean that you should ignore this metric? The number of employees at your top contract manufacturer who are working more than 60 hours per week is increasing – does this mean you should raise an alarm?

The most common method for analyzing the data is to track trends against previous time periods and/or make comparisons against a goal or control limit. Another effective method is comparing the individual facility records against one another. Comparing performance across internal entities (e.g., safety performance by facility or business unit) or comparing your company's suppliers (e.g., violations of the 60 hours per week limit by a supplier factory) will inevitably focus attention on the entity with the worst score. Even if the overall trend is good, as in the safety example, the competition between facilities could drive even greater performance.

If you are aware and prepared for the consequences of this presentation, it can be incredibly effective. The poorly scoring entity has an urgent imperative to understand and explain why its data is poor and what it plans to do to address the situation. This automatically results in the management of that entity taking ownership and prioritizing the issues. In my experience, this method works best when you take the time to notify the lower-scoring entity in advance and patiently work through the data with them. The last thing you want to do is to point the finger at someone based on spurious data. Once the data is validated and root causes are understood, you can turn ire into respect if you take the time to work with the managers on an action plan. I consider it poor form to surprise a manager with poor results in an ops review without a preview. A good rule of thumb for ops reviews is "surprises are only fun at Christmas and birthdays."

Once you feel that there are enough data to constitute a "disturbing trend," your responsibility is to understand the root cause of the problem and assemble an action plan to address this issue that involves the essential stakeholders. Don't just present the data and expect the remedy to emerge from the discussion. Doing this not only indicates that you are unprepared, but it often results in a random set of actions that may not be all that helpful. Expect that you will get feedback and perhaps a change in direction on the action plan you propose, but that is far more preferable than making up a plan during the meeting. Make sure that you follow up with regular progress reports on your action plan until the KPIs return to an expected trend.

Running effective management committees

As discussed above, corporate responsibility touches just about every aspect of the company; yet the corporate responsibility program has authority over very few of these issues. To ensure that the stakeholders with authority over corporate responsibility issues are aligned on a single strategy with consistent goals and measures, many companies assemble a committee of decision-makers from each of the functions covered by the corporate responsibility program. Intel called this the management review committee (MRC), and at AMD it is the corporate responsibility council (CRC), whereas Apple operates without formal committees.[53] Since many companies operate via cross-functional groups, below are steps for establishing and operating effective committees:

Step 1: Selection of the membership

The reactions to an invitation to participate on the corporate responsibility committee can range from "run-and-hide" avoidance to competition over who will be the representative. The best practice from my experience is to discuss the purpose and goals of the committee with the executives in charge of each represented function, and ask them to formally designate a representative. In this discussion, you should emphasize that the representative should be someone with decision authority and visibility within the organization. Equally important is that the person has interest and passion for corporate responsibility. For example, in addition to my role as facilitator, the AMD Corporate Responsibility Council includes:

- Brand marketing VP

- Information technology VP

- Public affairs VP

- Executive office VP

- Communications VP

- Ethics and labor law associate general counsel

- Strategy director

- Product group director

- Technology group director

- Legal director

- Human resources director

- Diversity and inclusion director

- Sales group director

- Government affairs director

Step 2: Gaining commitment

Once the membership has been appointed, it is wise to set up one-on-one meetings with each of them to build a relationship and learn a bit more about their interests and expectations for the group. I have found that shared ownership is the most effective means to running a healthy committee. By distributing responsibility for outcomes to the appropriate committee representatives, everyone becomes invested in the process. For example, if one of your goals is improving the diversity of your workforce, the representative from HR should be tasked with reporting on the goals, progress, challenges, and plans for this function at regular intervals.

Another effective management technique is to ensure that each representative on the committee has incorporated specific and measurable goals into their individual performance plan and/or departmental goals. In other words, effective participation in the committee will be graded as a part of each representative's performance review. Some companies have taken this a step further by establishing corporate responsibility goals into compensation formulas; for example, a business unit manager would receive a better bonus payout if safety targets were achieved. Pay and performance incentives are incredibly effective means of driving program goals.

Step 3: Establish standard processes

The cadence of your management committee meetings depends on the maturity of the program. The less mature the program, the more often the committee should meet. For example, early-stage programs will need significant meeting time to work through the materiality and strategy-setting processes outlined in the previous chapter. Quarterly operational reviews are appropriate for more mature programs. In these cases, the strategy, goals, and responsibilities are well under-stood and the group comes together mainly to review performance. The minimum frequency for an effective team is quarterly. Longer duration between meetings results in members forgetting the elements of the program and becoming less engaged.

As with all group meetings, there is no substitute for effective prep-aration. I like to establish an annual plan for the group based on mutu-ally established goals. You should establish a recurring schedule for each of the goal owners to report to the group on their progress and plans. Before each meeting, publish an agenda and the expected out-comes of the meeting (especially if decisions will be required) and distribute pre-reading materials in advance. This will allow partici-pants to prepare with their management and engage decisively in the meeting.

Step 4: Running the meetings

An essential element to running effective committee meetings is to act as a facilitator rather than presenter-in-chief. By distributing the speaking roles to others you simultaneously make the meeting more interesting and ensure engagement of others. If you take on the bulk of the speaking roles, the meeting can start to resemble a personal agenda, which will disengage your audience. Multiple presenters on the agenda implies broader engagement, investment, and ownership. A few basic tips for running an effective meeting include:

- Whenever possible, hold your meetings face-to-face. The level of interaction and discussion will always be richer and more productive

- Send the materials in advance over e-mail, or even send printed copies to those who are not great on e-mail (sometimes I will use multiple means to distribute materials such as e-mail, print, and inserting into the meeting notice). The point is that your participant should be prepared for the discussion rather than making snap decisions

- Schedule enough time for discussion for items on the agenda. The focus should not be to get through all the PowerPoint slides. The entire reason to bring the group together is to tap into the collective brainpower and varied perspectives of the group. If all you accomplish is a presenting a slide deck, you have squandered the capabilities of the team and potentially turned off some of your participants

- Manage the administrative details well. I start every meeting by taking attendance and recording this in a table that is sent with the minutes – it is amazing how this small step increases attendance. I then read off the action items recorded from the last meeting, and we determine as a group if these were accomplished or not. I maintain a single document to record attendance, action items, decisions, and the minutes in bullet point format. After each meeting, I update the document and send it to the members within 48 hours. Because people will forget their action items between meetings, I make sure that each action owner is reminded a few days before the meeting.

Step 5: Recognition matters

Another means of running an effective management committee is recognition. It is a basic truth that we all respond to recognition. By recognizing the people who contribute to the process you not only reward them for their work, you also motivate others to step up their participation. Be careful not to overdo recognition; if everyone gets an "Attaboy!" it becomes trite and can backfire.

Continuous improvement: Focused but flexible

As discussed in the previous chapter, it is essential to focus your program on the issues of greatest importance to your company and society. Once these "material issues" have been established and you have a mature system to set goals and track performance on these issues, your work is not over. The hallmark of world-class corporate responsibility programs is striving for continuous improvement.

For example, your program may have achieved awards for excellence in employee diversity and inclusion. If so, congratulations . . . now what more can you do? Effective managers are always looking for that next increment of improvement, the innovative ways to define leadership best practices. In other words, corporate responsibility is a journey, not a destination. Not only are there more actions you could take to improve on the issues you currently track, but the field is still rapidly expanding and changing. There is no time to rest on your laurels; a good manager is always looking ahead for ways to improve.

There is a difficult dichotomy inherent in the continuous improvement doctrine, that is: being focused versus being flexible. Data-driven programs are laser-focused on driving performance improvement on the issues that are material to the business. They actively disinvest in tangential issues that could be distractions and dilute their focus. The tough part, the part that requires judgment, is balancing the need to focus with the flexibility it takes to identify material changes that impact your program.

Each year, there are new challenges and new issues that can affect your corporate responsibility program. The methods to identify these emerging issues are discussed in Chapter 3. Part of your day-to-day responsibilities, regardless of where you stand in the corporate hierarchy, is to find and assess these emerging issues. If you judge a new issue as a significant threat or opportunity to your program, it is essential to adapt. Any worthwhile strategy is just a blueprint, a living document that must change as conditions dictate. As you implement your corporate responsibility program, anticipate change and adjust your approach accordingly.

As discussed in Chapter 2, essential characteristics for corporate responsibility managers include curiosity and the capacity and desire to learn new things. Good corporate responsibility managers are eager and excited to understand new challenges and opportunities, analyze situations, and take well-considered risks to advance their programs.

A recent example demonstrating the need for flexibility is the relatively new issue of "conflict minerals." In a very short time, the presence of conflict minerals in the supply chain for many products has emerged as a top-priority corporate responsibility issue. Much like the theme of the movie *Blood Diamond*, the issue connects the uses of metals in consumer products (such as cell phones and computers) with bloody conflict and human rights abuses funded by the profits from the extraction and trade of these materials in Central Africa, particularly in the Democratic Republic of Congo.

Having worked for years in multinational electronics firms, never in my career did I think I would be focused on mining issues in Central Africa. The standard reaction to issues like this used to be that they were too far removed – too many steps down the supply chain – for most companies to get involved. But, because of the scale and seriousness of the human rights abuses, the fact that activists have effectively linked these issues to the electronics sector, along with the recent enactment of the Dodd–Frank law requiring due diligence for sourcing these materials, the issue is now at the top of my priority list. While this is a great example of an emerging issue, it also could be one of the most important issues I will ever work on if we are successful in eliminating or reducing the truly egregious abuses in the mining fields of Central Africa.

5
Environmental sustainability

This chapter defines the structure, functions, and typical goals of a corporate environmental department at a high level and outlines methods for integrating environmental management with corporate responsibility.

> Keep close to Nature's heart . . . and break clear away, once in a while, and climb a mountain or spend a week in the woods. Wash your spirit clean (John Muir).

Similar to my own experience, many people who migrate into the field of corporate responsibility have a background in the environmental field. For these people, this chapter will be too basic. The concept underlying this chapter is to provide the information needed for the corporate responsibility manager to effectively interface with the company's environmental team. So, rather than delving into all of the details around environmental management, this chapter will establish a framework and provide practical tips for integrating environmental sustainability into corporate responsibility.

The term "sustainability" has become synonymous with environmental issues, but its roots are much broader. As defined in Chapter 1, environmental, social, and economic issues are the three pillars of sustainable development – the so-called triple bottom line – also known by the alliterative phrase, "people, planet, profit." In practice, however, most people associate the word sustainability with environmental protection. Let's start this chapter by identifying where the people with control over your company's environmental footprint live.

Where environmental groups live

Depending on the business model of your company, environmental management responsibilities are typically split between these four business units:

Manufacturing

This is the most common home for environmental professionals. The primary responsibility for these folks is compliance with environmental laws and regulations. In many companies, this role has expanded to include cataloging and managing the company's "environmental footprint."[54]

The manufacturing group is the natural home for environmental professionals because manufacturing facilities tend to have disproportionately large environmental impacts and regulatory liabilities. In many companies, this discipline is housed together with the health and safety disciplines into the environment, health, and safety department (EHS). The people in the environmental department usually have technical backgrounds and fairly tactical responsibilities. For example, your company's environmental staff might fill out permit applications, collect and record the data from pollution control devices, or calculate your company's carbon footprint.

Facilities and real estate

As the name suggests, this team is responsible for all of your company's buildings and building services. While their level of expertise is not as technical compared with the manufacturing group, the people in this group control much of your company's environmental performance. For example, the facilities group is usually responsible for buying or building new structures and thus controls green building standards such as Leadership in Energy and Environmental Design (LEED). This group also controls building services such as utilities (including green energy), heating and cooling, water use, lighting, landscaping, cafeteria maintenance, transportation, and other functions that have a significant impact on environmental performance.

Product design and quality

As the focus of environmental management has expanded, stakeholders are increasingly focused on the impacts of your company's products. For example, regulations to ban lead and other toxic materials in electronics, mandatory take-back laws for products, and restrictions on product packaging are all focused on products as opposed to the factories that made the products. Along with these new regulations, there has been an increased corporate focus on "green marketing" of products covering everything from energy efficiency to reduced packaging.

In some companies the people who deal with product-environmental issues are grouped together under a "product stewardship" department. Most often these responsibilities are divided between the engineering (environmental product design), product quality departments (regulatory compliance), and marketing (green advertising).

Purchasing, procurement, or supply chain

As companies continue to outsource the production and manufacturing of their products, the environmental footprint of the supply chain has become a major focus. For example, in the area of climate protection, many companies are accounting for the climate impacts of their suppliers in their overall carbon footprint.

Some companies manage the environmental impacts of their suppliers within the procurement organization, while others hand this off to the facilities or manufacturing environmental department. Often the procurement and EHS departments form a partnership on these issues because the procurement team owns the relationships with suppliers and the EHS team has expertise on environmental issues. It is important to note that many companies have yet to embark on a program to account for and/or manage their suppliers' environmental impact. We explore the area of supplier responsibility in the next two chapters.

Another environmental management role within your company's purchasing department deals with "green procurement." Examples include everything from buying recycled paper to organic food offerings in the company cafeteria. In addition, the purchasing department may control the consumption of basic resources such as water, electricity, or raw materials. The people in these jobs can have a pivotal role in any conservation effort or in establishing the specifications for purchasing of environmentally preferable products.

There are a wide variety of roles within a company that have significant leverage on your overall environmental footprint. Within each of these departments there can be a number of sub-specialties. For example, EHS departments often include experts on air quality, climate protection, waste management (hazardous and non-hazardous), water quality, water conservation, wastewater treatment, pollution control, pollution prevention, energy management, and others. Each of these roles can be highly technical, and the people working in these jobs are experts in the regulatory requirements, engineering, and equipment associated with their niche.

Engaging on environmental issues

To engage on environmental issues, the first step for the CR manager is to understand the organizational structure, roles, and responsibilities of the environmental staff. Similar to the other program areas under the CR umbrella (e.g., diversity, ethics, etc.), your role is to represent the accomplishments and plans of this group in your CR communications and to form working partnerships to influence improved performance. As you engage with your environmental team, there are a few overarching tips that can serve as a guide for the partnership between corporate responsibility and environmental management:

Match your effort to the opportunity

Each company's environmental footprint is different. Work with your environmental staff to understand the scope and distribution of your company's environmental impacts. This inventory should serve as a guide for you to prioritize where to invest your time and effort. For example, when AMD transferred major manufacturing assets to a joint venture, the company's carbon footprint decreased approximately 75%. Until the transfer, these manufacturing plants were the primary focus of the corporate responsibility program. Today, with the factories no longer in the mix, AMD is focusing on the environmental impacts from its suppliers and products; a terrific segue into the next topic.

Think lifecycle

Sometimes the biggest environmental impacts are not the most apparent. "Lifecycle assessment" is a process that examines all of the environmental impacts of a product from raw materials through manufacturing/production, distribution, consumer use, and final disposal. This kind of analysis often produces surprising results. For example, Levi's conducted a lifecycle assessment of its jeans, which demonstrated that the majority of the environmental impact came from washing the jeans after they were sold. This finding broadened the company's priorities. While Levi's continued to make improvements in the manufacturing process, after the lifecycle assessment it began to invest in new programs to reduce post-sale impacts by encouraging customers to wash their jeans less often, wash in cold water, or with fewer detergents.

A similar example comes from my own experience at AMD. When the company looked at the lifecycle carbon footprint of one of its new computer chips, the study showed that about 80% of the overall carbon emissions stemmed from the use of energy by the chip after it was installed in a computer. Like the Levi's case, AMD continues to manage the manufacturing-related environmental impacts, but this finding raised awareness of the importance of mitigating the energy consumed (and carbon generated) by a computer chip in use.

These examples illustrate that the bulk of the environmental impact of a product or process can be outside of your company's direct control. While Levi's controls how the jeans are made, after they are sold it is the customer who decides how to wash them. The techniques for managing issues outside of your company's direct control can range from a redesign of the product to raising consumer awareness. In the Levi's case, the company partnered with Whirlpool Corporation (makers of high-efficiency laundry machines) to increase public awareness about how to wash clothing with less environmental impact. In the AMD case, the company is designing energy-efficient chips for real-world workloads and continuing to work on ways to better manage the power used by PCs, servers, and game consoles.

Engage employees

In recent years, many large and small companies have started green teams. These eco-minded volunteers are focused on helping the company become more sustainable. I have helped launch three different green teams at Intel, Apple, and AMD. Each of these teams took on similar roles, such as reducing cafeteria waste, increasing recycling, and conserving energy and water. As a corporate responsibility manager, you may be called on to launch or support your company's green team. You might also be in the role of liaison between your green team and your facilities department because they often propose ideas that require funding or approval. By supporting and nurturing your company's green team, you can build a "reserve army" of talented people to help you achieve your company's environmental goals. Also, as will be discussed in Chapter 11, green teams are correlated to overall employee satisfaction and engagement.

Motivating green

An example of how to unlock employees' energy to improve the environment comes from my work at Intel. A couple of years into my job as the corporate environmental manager, I started a program called the environmental excellence award. The idea was to recognize employees for their individual and team contributions to improve the environment. We worked with the employee communications team to publicize the nominations process across the entire company and the response was amazing. We had far more nominations than expected, making selection of the winners a tough job. A senior executive presented the awards and the coverage of this event went out to all 100,000 employees. Nominations continued to soar in subsequent years, and soon we had to establish more award categories and a team of judges (who were recruited from the green team). Employees in every department submitted projects, ranging from personal initiatives like bicycle commuting, to corporate-level actions such as reducing the use of hazardous chemicals in manufacturing processes. Over the following years, this award was a significant engine of environmental improvement for the company, all stemming from a grassroots effort.

Environmental goals

Because environmental issues are important to the public, this area is often the primary metric used to judge the effectiveness of your corporate responsibility program. If the environmental team in your company has not established goals and/or systems to track their progress on carbon, water, or other environmental indicators, setting goals should be the first thing you work on with them.

Start the process by establishing your company's current baseline of pollution and resource consumption. Once you have the baseline in place, you can work with the EHS team to establish improvement goals for each of the priority environmental issues. Typically, the EHS team will forecast the resources required, and will have a reasonable degree of certainty that they can achieve the goals, before announcing a public commitment.

If it is not possible to set quantitative goals for your company's environmental performance, establishing qualitative goals is a good alternative. For example, if the company has not yet established systems to track carbon emissions, you could commit to filing a carbon footprint analysis with the Carbon Disclosure Project (www.cdproject.net) by a certain date.

There is a wide variability in corporate environmental goals, ranging from pollution prevention, hazardous waste reduction, recycling, land conservation, protecting underground water sources, and many others. Your company's goals should correspond to its top environmental impacts. Below are a few environmental goals that are often reported by companies:

Carbon footprint

One of the must-have goals for any modern environmental program is reduction of your carbon footprint. Inherent in the development of a carbon reduction goal is the development of a full carbon inventory (or footprint) for your company. There are three categories of greenhouse gas emissions that will be tallied in your company's carbon footprint:

- **Scope one:** Direct emissions of greenhouse gasses from your company's operations to the atmosphere

- **Scope two:** Indirect emissions created by the energy used by your company

- **Scope three:** Other indirect emissions stemming from activities such as logistics, travel, employee commuting, supply chain, etc.

Many company carbon footprint assessments are limited to only scope-one and scope-two emissions because it is difficult to accurately account for scope-three emissions. Once you understand your carbon footprint, the next step is to establish climate protection goals. Since most companies are focused on growing, carbon reduction goals are often expressed as a function of the company's production. These are called "intensity goals" because they are based on reducing the amount, or the intensity, of greenhouse gasses released for each unit of production. Some stakeholders are opposed to intensity-based goals because, even if the carbon per product is reduced, the overall emissions to the atmosphere could continue to increase as production grows.

In contrast to intensity goals, some companies have established absolute reduction targets. For example, the 2010 Coca-Cola Sustainability Report included an overall cap on the company's carbon footprint: "Grow our business but not our system-wide carbon emissions from our manufacturing operations through 2015, compared with a 2004 baseline." Climate protection goals can also be broken down into sub-goals that focus on particular aspects of the company's carbon footprint. AMD has established an intensity-based goal for its remaining manufacturing sites, as well as an absolute carbon reduction goal for non-manufacturing sites. In addition, some companies have made commitments to purchase renewable energy or buy carbon offsets[55] as a way to mitigate their carbon impact. For example, Starbucks established a goal to "purchase renewable energy equivalent to 50% of the electricity used in our company-owned stores by 2010." While offsetting your company's carbon emissions through green energy or carbon offsets is a fine strategy, it is not sufficient. Environmental stakeholders will hold your company accountable for its emissions and will demand to see significant progress in making absolute reductions.

Water footprint

There is an increasing awareness of freshwater scarcity as a top environmental issue for companies. The importance of water as an environmental concern for your company is proportional to the relative amount of water the business consumes. Even if your business is a small water consumer relative to other companies in your industry, you still need to understand your company's water usage and conservation goals. For example, if some of your company's facilities are in locations with scarce or fragile water supplies, water conservation will likely become a top priority in your corporate responsibility program regardless of your relative water consumption.

Like carbon, a water reduction goal starts with a good inventory of all the water used by your company. The goals can be either intensity-based or absolute use reduction. For example, AMD established an intensity-based goal of 20% water-use reduction per employee in non-manufacturing sites, whereas Starbucks has set an absolute water reduction goal to "reduce water consumption by 25% in our company-owned stores by 2015."

Waste goals

Another common corporate environmental goal is waste reduction and recycling. In contrast to the intensity-based format for carbon and water, waste goals are often expressed as a percentage of the waste that is diverted from landfill. Diversion goals are useful because they encompass all of the company's waste reduction and recycling efforts into one measure, which is the amount of waste kept out of landfills.

Perception is reality

Last, but certainly not least, it is critical to take into account stakeholder views and expectations when establishing your company's environmental priorities. The purists would prefer that environmental priorities be determined by a scientific assessment of risks or regulatory compliance requirements. In the real world, it is not so black and white. The outrage that external stakeholders may feel about the environmental risks posed by your company does not always track with scientific risk assessment. It is essential, however, that your program includes these perceptions in its decision-making and priority setting.

Peter Sandman is a leader in the field known as risk communication. The central thesis of risk communication is that the perceptions of stakeholders trump scientific calculations of risk.[56] While the environmental scientist will tell you that "Risk = the level of the hazard multiplied by the level of exposure," Sandman will tell you that "Risk = the perceived hazard multiplied by the level of public outrage."

What this means is that the perception of the threat dominates the discussion and must be factored into your program's response.[57] We won't cover the many reasons why the perception of risk does not reflect the scientifically calculated odds, but a simple example makes the point: People are very comfortable driving in cars, which carries a relatively high risk of death or serious injury, yet the same people would be extremely uncomfortable if they were told that the groundwater near their homes had been polluted even if they never drink the water.

Years ago, Dr. Sandman conducted a training seminar at Intel's New Mexico facility after consistent and escalating concerns over the impact of environmental emissions in the surrounding community. The message from this training was that, no matter how much data and scientific information Intel presented, the public near the facility would remain outraged. In fact, presenting more analyses and data might further enrage the public because they would perceive it as a smokescreen. It did not matter if the data showed the risk to be infinitesimally small; Dr. Sandman taught the group that reasonable people can, and often will, react based on their perception of risk, which can be orders of magnitude different than the calculated risk. Once the team understood this message, the dialogue with the community switched from an argument over whose view was more valid, to listening to the public and taking actions to mitigate their concerns.

Conclusion

Corporate environmental management is very a technical topic, the full breadth and depth of which is beyond the scope of this book. As with all the programs covered by your company's CR program, you need to understand the fundamentals but you do not need to be an expert. By understanding the structure, functions, people, and goals of your company's environmental program you should be able identify the big risks and opportunities and develop a plan to efficiently and effectively interact with this team.

6

Supplier responsibility:
Part 1
Establishing the program

This chapter sets the context for "supplier responsibility" and covers the steps to establish your program.

All is connected . . . no one thing can change by itself (Paul Hawken).

There are probably more jobs and more job growth for corporate tree-huggers working on supply-chain issues than in any other area of the company. Because of the rapid growth in this area, unlike the previous chapter on environmental sustainability, this chapter will delve into the details of how to set up and run a "supplier responsibility" program. For this area, the word "treehugger" may not be the most appropriate because, as you will see, most of the priority issues in this space tend to be social (i.e., labor and human rights).

Setting the context

Decades ago, companies figured out that it was far cheaper to out-source their manufacturing to suppliers in "low-cost" countries. While the primary driver behind outsourcing is cheaper labor rates, many of these low-cost locations suffer from lax labor and environmental regu-lations and/or poor enforcement. As a result, an increasing number of companies are hiring people to monitor conditions in the factories that manufacture their products.

The most recognizable of the low-cost manufacturing countries is China. I have spent a good deal of time in China, first with Intel, and then with Apple. At the time of writing, the current Chinese minimum wage is about $200 per month (this wage has been rapidly increasing and varies by location). Compare this to the U.S. minimum wage of $7.25 per hour ($1,160 per month for a 40-hour week without over-time) and the advantage is obvious: a Chinese worker will work for less than 20% of the cost of an American worker. In other popular outsourcing countries (e.g., Vietnam, Cambodia, Philippines), the dif-ferential in labor rates is even more skewed in favor of outsourcing. This trend, combined with improving quality, flexibility, speed, and logistics, has made outsourcing to China and other hotbeds of manu-facturing irresistible. The cost advantages of outsourcing are now so compelling that one could argue that it is the fiduciary responsibility of the company to use this strategy.

One more piece of context for the outsourcing trend is called the "smiling curve" (also called the "smiley curve"). Stan Shih, the founder of Acer (a computer company headquartered in Taiwan), first proposed the smiling curve in 1992.[58] The curve shown in Figure 3 is a simple illustration of the amount of value added throughout the product lifecycle. The reason that the curve is turned up on both ends – like a smile – is that the beginning and the end of the lifecycle com-mand higher returns on investment compared with the middle part – production. In other words, the profit margins for companies that are focused on product design or marketing are much higher than for companies focused solely on manufacturing.

Ultimately, the smiling curve makes intuitive sense: value is created from innovation. For most manufacturing processes, there is very little creativity or innovation involved. The goal is to optimize product quality against speed and cost – fairly standard stuff for most products (the exceptions to this rule are the products where manufacturing is incredibly complex, such as making semiconductors on silicon wafers).

Figure 3 The smiling curve
Source: Dedrick et al. 1999[59]

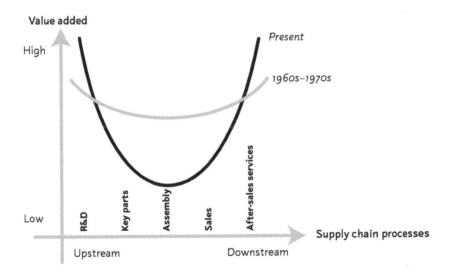

A smart manager reacts to the smiling curve by focusing the company's efforts on the activities that produce the highest value, or return on investment – design, marketing, and sales. This also means that the business manager's imperative is to cut costs in the middle part of the curve, assembly or manufacturing, where the margins are the slimmest. Cost-cutting almost always leads to outsourcing manufacturing operations to low-cost locations.

While simple, this observation is a very powerful insight because it is at the heart of the outsourcing movement. If the smiling curve was the spark that lit the outsourcing movement, then globalization was the gasoline. The advent of globalization – which relaxed trade barriers, made information instantly available, logistics cheaper, and international capital more accessible – made it easier for businesses of all types to use outsourcing. As shown in Figure 3, the smiling curve has become even more pronounced in recent years. With globalization, every manufacturing business could be linked to a foreign manufacturer who could produce their designs at a fraction of the cost.

So, how does this trend create jobs for corporate treehuggers? Since the mid-1990s, when Nike became the poster child for outsourcing its production to sweatshops in Asia, public attention and corporate awareness have been focused on conditions in the supplier's factories. The Nike story was quickly followed by related stories in the apparel industry such as child laborers making clothing promoted by Kathy Lee Gifford and the similar conditions in factories making clothes for Gap.

These high-profile cases changed everything. The reputations of companies on the right and left sides of the smiling curve that focused on design, branding, and sales, were suddenly at risk from the public outrage over poor treatment of their suppliers' assembly-line workers. Like a post-globalization industrial revolution, activists have successfully attached these issues to global brands, forcing them to drive improvements.

At first, Nike pushed back against the activists by saying that it did not control the companies that supplied its shoes. When it was clear that this strategy was tarnishing the entire brand, the company changed course and figured out ways that it could in fact control the practices at these factories through its buying power. Similarly, when made aware of conditions in Honduran factories making clothes with her label, Kathy Lee Gifford originally said she was not involved in the production of the clothes. Later, she contacted federal authorities to investigate the allegations and began supporting laws to protect children from working in sweatshop conditions. Gap assembled a global team of auditors – over 100 people strong – to stamp out poor conditions in its supply chain, which, as you will read below, can be a tough job.

Supplier responsibility in electronics

I first encountered the supplier responsibility issue when I worked at Intel. Intel was informed that a news organization was planning an exposé on one of the company's outsourced circuit board manufacturers in Hong Kong. Up to this point, Intel had adopted the prevalent attitude at that time that suppliers' behaviors were their own affair – Intel's involvement was limited to the product or service we buy from them.

Thinking that an accusation was imminent, Intel scrambled a team to visit the factory to assess the situation. There were a few gaps in their program but nothing on the scale of the apparel industry stories. Ultimately, the story never aired, but after this experience, Intel took a hard look at this issue, and decided we had to make some changes to address supplier behavior on labor and environmental issues. The program began with a training class for product quality teams – in essence, asking them to monitor conditions in supplier factories. The issue stayed quiet for several years in the electronics industry (while continuing to make headlines in the footwear and apparel industries) until the early 2000s. Around this time, the company started to observe significant increases in the number of inquiries from customers about Intel's practices. Many of these inquiries were focused on environmental issues in the aftermath of an incident involving Sony.

In December of 2001, the government of the Netherlands seized 1.3 million Sony PlayStation® game consoles. The estimated value of the items seized was $162 million. The reason for the seizure was that there was too much cadmium in some of the cables. Sony had outsourced much of the production of the PlayStation® and a supplier installed the cadmium-tainted cables. Given the very high cost of this episode in lost product and brand reputation, Sony started its "green partner" program. This program set out to audit every supplier involved in the PlayStation® and other Sony consumer products for compliance with Sony's code of conduct. Soon, other electronics firms followed suit and every electronics supplier experienced a huge increase in the number of customer surveys and audits focused on social and environmental issues.

During this period, I was the Director of Sustainable Development for Intel. Intel had joined the trade association, Business for Social Responsibility (BSR), and started a small group called the High Tech Coalition that included Sony and several other branded electronics firms. In early 2004, the group was meeting at Intel's offices in Chandler, Arizona when we collectively decided to take on supplier responsibility in the electronics industry as our primary focus. The goal was to harmonize the standards and assessment methods we would all use to hold each other accountable for social and environmental issues. From this small team and lofty mission, the Electronic Industry Citizenship Coalition (www.EICC.info) was born. We officially announced the new organization at the 2004 BSR conference in New York City with seven member companies. Today there are more than 70 member companies in the EICC.

As the group struggled to agree on the specific standards to use within the electronics industry, it appeared that the effort might founder. Two things changed that brought the EICC to fruition:

- **Birth of the Electronic Industry Code of Conduct.** During this time, Dell, HP, and IBM had been in negotiations with several NGOs over shareholder proxy resolutions focused on conditions in their supply chains. These negotiations produced an agreement on a five-page "Code of Conduct." Because this Code was the product of a multi-stakeholder agreement, the other companies in the coalition quickly approved it and it became the first version of the Electronic Industry Code of Conduct

- **The iPod story.** During this same time period, poor working conditions at a Chinese factory making the iPod were highlighted in a British tabloid, the *Daily Mail*, and the story quickly went viral. Suddenly, the industry had a "Nike problem," with the hottest brand in the industry being called out for poor conditions in its supply chain. With this exposé, interest in the EICC soared. Apple quickly joined the group, as did a flood of other companies from across the electronics supply chain

Genesis of a supplier responsibility program

Soon after the allegations of sweatshop conditions in Chinese iPod factories, I accepted the role to create and run Apple's supplier responsibility program. Apple was under tremendous pressure to take action and address the allegations. The scrutiny was so intense that Apple posted the job announcement for the new position for only one day because it had attracted press inquiries.

When I got to Apple, the management had already completed an audit of the factory that had been named in the press allegations as well as three others. For context, these facilities are massive; the largest has an employee population near 400,000. Apple had also worked with the leaders of these factories to address the audit findings and then issued a very candid statement that some of the allegations were true – such as long hours and poor living conditions.

On my first day, my manager handed me the four audit reports (50-plus pages each) and asked me to extract and categorize the major findings so that Apple could systematically ensure that each issue was resolved. Looking back, we were all a bit naïve about the depth and breadth of the issues we would have to tackle. There were many times that I felt overwhelmed and woefully under-resourced to deal with the social and environmental conditions in such a giant and complex supply chain.

Nonetheless, I distilled the findings from the four reports into high-, medium-, and low-priority categories and briefed the senior management on the summary. The picture was a mixed bag. The conditions were not as bad as described by the press and activists, but clearly there were areas for improvement. Soon, I was on my way to China to speak to the leaders of these four factories (a trip I would repeat at least once a quarter for the time I was at Apple). As you might imagine, these meetings could be contentious. In the end, the Apple team had the two ingredients that were needed to make a program like this successful:

- **Strong executive support**. The support for this program at Apple extended through the highest levels of the company

- **A strong market position**. Most companies are eager to work with Apple because of its large buying power and growth potential. Many suppliers were willing to take whatever actions were necessary (within reason) to secure the Apple business

After the initial crisis and intense scrutiny had eased a bit, I began the process of setting up the management systems for the program and hiring a team. There are many ways to establish a supplier responsibility program, but this chapter follows the general approach I took at Apple. While some program elements may not apply to your company, this outline can serve as a resource for you to pick and choose the attributes that fit your situation best.

A word of caution about the world of supplier responsibility: if the two attributes mentioned above – executive support and market power – are not in place or are weak within your company, you will likely need to make some compromises. Ultimately, this field is about trying to get people from other countries and cultures to conform to international standards and norms. This is a tough job in any situation, but it is a whole lot easier when you have the backing from your company and enough money flowing to the suppliers to get their attention.

Establishing your company's code of conduct

The first question you should be able to answer is: Why do you need a code of conduct? The answer to this lies in the fact that most companies now have a global supply chain. The applicable laws and regulations for social and environmental issues vary from country to country. Most contracts already contain boilerplate stipulations that the supplier must comply with all applicable laws and regulations. But if the supplier is in a country where labor and environmental laws are inadequate or poorly enforced, the purchasing company could be exposed to liability for using sweatshops. In essence, the code of conduct establishes a consistent "floor," or the minimum expectations for your suppliers, regardless of the locally applicable laws or enforcement.

If your company does not have a formal supplier code of conduct, start by benchmarking with others in your industry as well as reviewing your company's own internal policies, such as business ethics, human resources policies, and EHS policies. There are also international standards for many aspects of social and environmental performance. For example, Social Accountability 8000 (SA8000)[60] is a highly regarded standard for labor rights which sets out international norms for maximum working hours per week, minimum days of rest per week, the definition of child labor, and the right to collectively bargain with management, as well as other issues.

Even if your company is in another industry sector, I would recommend starting with the EICC, because it is based on internationally accepted standards and it is comprehensive. The EICC covers five main areas of conduct:

1. Labor and human rights

2. Environment

3. Health and safety

4. Ethics

5. Management systems

Whether you start from the EICC or another code, you will need to get executive support. The best way to approach this is to gather the codes of conduct from your benchmarking analysis and your company's existing workplace policies, as well as any applicable international standards, into a "side-by-side" analysis. For each issue, list the policies for the company codes that you benchmarked, any relevant policy from your own company, and the applicable international standard. Based on this analysis, you should be able to quickly determine if your company has any gaps compared with other codes and standards.

Cut and paste the verbatim language from each source into the table shown in Figure 4 and compare them word for word. While this is a useful analytical tool, it is not necessarily the best tool for communication to senior management. With limited time and attention spans, senior management typically wants you to get to the bottom

Figure 4 Side-by-side analysis for development of a code of conduct

Section	Issue	Your company's requirement	Benchmark company requirement	International standard	Recommendation
Labor and human rights	Discrimination				
	Child labor				
	Freedom of association				

line as quickly as possible. With this in mind, use a table similar to Figure 4 as an analytical tool and backup reference for your summary of the recommendations for developing your company's supplier code of conduct.

The analysis is similar whether you are starting from no code or trying to improve an existing code. In essence it boils down to your recommendations on the standards that your company will establish for its suppliers and the rationale for your recommendation. For example, on the issue of child labor, SA8000 sets the minimum working age at 15 years with exceptions for qualified apprentice programs. The EICC also uses age 15 as the minimum working age but respects the applicable country law if it is higher than 15 (for example, the minimum working age is 16 in China). Your recommendation could be to adopt the child labor definition from EICC or SA8000, or a modification based on your company's existing policies. In any case, the side-by-side table will allow you to reference the sources for the standards in your code.

In earlier chapters, we talked about "reading the system," or understanding the political paradigm and reacting appropriately. When recommending a new policy for your company, you will need to understand your company's position on the overall issue of supplier responsibility. For example, Apple was motivated to establish a leadership code based on clear, actionable standards. The company started with the EICC code and made changes that both clarified and strengthened the requirements. In contrast, other companies may choose to take a "light touch" approach by establishing only broad principles rather than a code of conduct that explicitly states the expectations for each area. For example, the Institute for Supply Management's Principles of Sustainability and Social Responsibility outlines a series of high-level principles but stops short of setting specific standards for supplier behavior.[61]

Wherever your company ends up on this spectrum, it is essential that a written policy is established and ratified by the highest level appropriate in the firm. This is important because the policy is the only tangible indication of your company's expectations. If there is no written policy, there will always be a credible argument that the expectations were not clear, and thus any performance from your suppliers could be deemed acceptable.

Laying down the law: Put the code in your contracts

In my opinion, *the best practice is to include your company's code in supplier contracts.* It is important to have the code embedded in contracts, because compliance with the code will then be contractually binding for the supplier. The clause should contain language that provides your company with the right to audit the facility for conformance with the code and obligates suppliers to correct deficiencies in a timely manner. The provision should also state the possible remedies for breach (violating the code of conduct) that should include contract termination.

Apple reopened every supplier contract to add in a clause obligating the supplier to conform to the code of conduct and updated the standard contract with this clause for all new suppliers. When this clause was added into contracts, there were two camps of suppliers: those who read it and those who signed it. While this may sound flippant, many suppliers will sign almost anything to get the business. This is problematic, because it means that they have not done an assessment of their ability to comply with the new contractual requirements, and thus are unaware if they are able to comply or not. The suppliers that read the contract language and assessed their programs would often push back and negotiate for better terms, such as: advance notification before an audit; granting more time to correct deficiencies; or limiting the damages in the event of a breach. Apple spent months working through these negotiations and ended up with an array of individual agreements with suppliers. In each case, the core terms Apple insisted on were: (1) that the supplier had to adopt Apple's code or its own code if it was deemed equivalent; (2) suppliers had to allow inspections (audits); and (3) suppliers had to commit to correcting deficiencies. In retrospect, starting the program with this contractual work was essential to the success of the program, because later on as Apple audited these facilities, there was little room to argue about the suppliers' commitments or the standards they had agreed to meet.

Choosing which suppliers to monitor

As you set out the contractual conditions for your company's code, one of the first issues you will encounter is which suppliers should be covered. This can be a thorny issue depending on your company management's level of support for supplier responsibility. Start with the premise that it does not make sense to cover everyone. You can't expect to monitor code conformance for every supplier who does business with your company. For example, your office supply vendor, your transportation company, or the landscaping company that trims the grass around your office are unlikely targets for code-conformance monitoring. These types of supplier are typically called "indirect" suppliers because they provide a product or service that is not directly related to your company's product or service. Most supplier responsibility programs focus on direct suppliers – those companies directly involved in your company's product or service. Establishing priorities for monitoring suppliers is critically important and involves a series of judgments outlined in the steps below:

Step 1: Create your initial target list

Apply some general criteria to filter your company's overall supplier list. Broadly speaking, the object of this exercise is to prioritize the suppliers that are closest to your company's business model and, by extension, your company should be able to exercise significant influence to monitor the suppliers' behavior. Another way to look at this exercise is to identify the suppliers that could do the most harm to your company's brand if they are committing serious code violations. There are two main criteria to determine which suppliers should be covered by your supplier oversight efforts:

- Amount of spending (this is often referred to as "spend" and expressed as a percentage of the overall spending on all suppliers) and/or the strategic value of the supplier (these could be two distinct criteria but are lumped together here because they both signify the importance of the supplier to your company)

- The likelihood that the supplier will violate the code based on generic data such as location of the supplier and/or the type of business

The most common way to identify suppliers that fall within the scope of your monitoring program is to focus on the direct suppliers that constitute 80% of your company's total spend (direct spending only). Once you have this list, then it is appropriate to conduct a "sanity check" with your supply chain management team. This review can identify any suppliers that should be removed from the list (e.g., perhaps the supplier will be phased out) and any that should be added (e.g., a new supplier that you will be ramping up significantly or is of strong strategic importance to the company). Next, you should look at the risk of code violations posed by these suppliers. The EICC has two risk assessment tools on its website (www.eicc.info), but there are many others and you can just as easily construct your own risk tool.

To assess the potential risk of your suppliers violating your code of conduct, the first thing you have to do is to identify the specific facilities that you are buying from. You may discover that your buyers only deal with the suppliers' sales people and may have no idea where the factories that produced the goods your company is buying are located. In this case, you will have to identify the specific facilities that supply your company. Once you have catalogued the facilities and locations, you can match up these locations against the lists of countries with known labor, human rights, or corruption problems published by Freedomhouse (www.freedomhouse.org) and Maplecroft (www.maplecroft.com) as a first approximation of the risk of violations that these facilities may pose.

Step 2: Refine your risk assessment

The next step is to refine the risk assessment process to narrow the list of supplier facilities that your program will monitor. Look at the type of work conducted by the suppliers. For example, assume that there is more risk of labor violations if the company has a large unskilled labor force with on-site worker dormitories. You can also assume that suppliers which utilize manufacturing processes entailing physical, chemical, or electrical hazards pose a heightened risk for occupational safety violations.

Step 3: Consider other known risks

Further refine your risk assessment by identifying other risk factors that may warrant scrutiny. For example, Apple discovered that suppliers located in certain countries had a higher likelihood of employing bonded laborers. Bonded labor is where workers have paid an exorbitant fee to get their job and are primarily working to pay their debt. The countries that have recently become more affluent typically import a lot of contract labor to fill lower-paid, unskilled positions that locals would not accept. In some cases, unscrupulous labor brokers will deliver workers to supplier factories who owe them (are bonded) for recruiting fees and expenses. This is a heinous practice that was found in factories with depressing frequency. After learning how to identify this practice, by insisting on interviewing contract laborers, Apple quickly figured out that the pattern was repeated in certain countries and identified these locations as an additional risk factor.

Step 4: Define the coverage of your program

Consider how many layers of the supply chain your program will cover. Most companies start their program with the suppliers with whom they have a direct relationship (tier one suppliers), but it is becoming more common to see companies diving deeper into their supply chains (tier two and beyond) to monitor code conformance. These are the suppliers to your suppliers. It becomes very difficult to approach these companies about code conformance since your company does not conduct any direct business with these suppliers.

The decision to approach tier-two+ suppliers is dependent on the company culture and the relationship to these suppliers. A typical approach is to "pass down" supplier responsibility from your direct suppliers to suppliers further down the chain. In other words, part of your conformance program is to ensure that your direct suppliers have a robust supplier responsibility program to monitor their own suppliers. Apple made the decision to monitor tier-two+ companies based on their strategic relationship. For example, if Apple directly engaged with the supplier in defining the specifications of their products, as opposed to ordering a generic commodity, then they were included in the monitoring program.

Step 5: Ask the targeted suppliers for a self-assessment

Send out a self-assessment questionnaire to the suppliers identified in steps one through four. In my experience, the self-assessment surveys can provide terrific insights into the day-to-day activities at supplier factories. The EICC has an excellent self-assessment questionnaire that delves into every aspect of its code. While self-assessments are not as reliable as audits, the results can be revealing. After reading hundreds of these, the responses range from programs that were clearly world-class to others that essentially self-declared that they had serious problems.

Be aware that it is a huge amount of work to ask your suppliers to fill out a self-assessment (for example, the EICC questionnaire is longer than 50 pages) and even more work for you and your team to evaluate the results. If you are going to use a self-assessment tool, start with a clear idea of how the data will be used. For example, would a poor score on the self-assessment lead to a supplier being put on the top of the list for an on-site audit? Does a good score mean that the facility would not be audited? The one item that is strongly recommended for any self-assessment process is to clearly indicate that the data will be verified if the facility is audited. Including this statement tends to increase the level of accuracy you can expect from the responses.

Step 6: Select suppliers for audits

Create a prioritized list of suppliers for your audit program. This task is often a mix of objective criteria (e.g., direct spend and facility locations) and subjective judgment (e.g., strategic value). This list is critically important in that it defines the scope of the audit program and the resources required. This is not a static list. You should set up a schedule by which you will review and revise your list of facilities to be audited.

Make sure that you have a process to evaluate any new suppliers that your company may hire. It is ideal to include a baseline audit or at least a self-assessment questionnaire as part of the supplier selection process. It is far more efficient and effective to identify and manage social and environmental issues as part of the supplier selection process, as opposed to doing so after the business relationship is firmly established.

This chapter covered the fundamental decisions, data requirements, and steps needed to establish a supplier responsibility program. In the next chapter we will cover the essential elements for implementing the program.

7
Supplier responsibility: Part 2
The four essential program elements

This chapter sets out four elements for implementing your supplier responsibility program and provides tips to balance compliance activities with other techniques for maximum effectiveness.

Trust, but verify (Ronald Reagan).

In setting up the Apple supplier responsibility program, I spent hours on benchmarking and research to understand how other companies (most in the footwear and apparel industries, which have years of experience in dealing with these issues) had structured their supplier responsibility programs. The goal of this research was not to copy what others had done, but to assess what had worked and what had not. Based on the results of this study, I established the following four program elements: compliance; business integration; capacity building; and communications and stakeholder relations.

Program element 1: Compliance

Most supplier responsibility programs are compliance-focused, with most of their resources consumed by conducting audits, correcting deficiencies, and following up. This chain of events sets up a paradigm where the customer is in the role of enforcer and the supplier is being regulated. The compliance paradigm is *necessary, but it is not sufficient for creating an effective supplier responsibility program*. As the next few sections will outline, additional program elements are equally important and you must titrate the emphasis that you apply to these programs to find the right balance for your company's goals. But first, let's explore the aspects of a robust compliance program:

The audit protocol

A good audit program is based on a *specific, objective, and actionable* audit protocol. The audit protocol is like a detailed instruction manual that instructs your auditors on *what* they should be auditing and the documents, interviews, and inspections they must review to determine conformance with the standards in your code.

Creating an audit protocol can be a laborious task. Start from your code of conduct and break it down into specific and actionable standards that can be audited. In some cases this is straightforward: for example, the EICC code states that no worker may exceed 60 work hours per week and must have at least one day off every seven days. While these are seemingly objective standards, even here you will need to define the meaning of a "work week" and set out specific formulas to evaluate the 60 hours per week and one day off in seven days standards. For example, you could state that a worker must have one day off for *any* seven-day period, or you could state that a worker must have one day off for a workweek defined as Sunday to Sunday. In the first case, the worker must always get one day off in any seven-day period, in the second case a worker could work for 13 days without a day off by getting the first and last day off in a 14-day period.

When you start developing your audit protocol, you will discover that your code is full of less objective standards, such as this example from the EICC code: "Worker dormitories provided by the supplier or a labor agent must be clean and safe and provide adequate emergency egress, adequate heat and ventilation, and reasonable personal space."

When an auditor is confronted with words like "reasonable personal space," it is best to put some numbers around the concept so that you get consistent and actionable results. Consider developing guidance documents that clarify each of the ambiguous areas of the code. In this example, your guidance document could set a minimum of 2 square meters per person in a dorm room, double bunks only (no triple bunks, no bunk sharing), adequate and separate arrangements for hanging clothing and storing personal items for each person, and so on.

By defining unambiguous standards for each area in your code, you can create actionable and auditable standards that your auditors can objectively assess. When you march through this process, do some research – you will find that other companies have defined many of these areas and you can choose the definitions that work best for your program. In addition, once you have developed your audit protocol with these supplemental definitions and guidance, it is best to share this information with your suppliers and/or hold training sessions so that they understand your expectations.

While it is impossible to eliminate all ambiguity from your audit protocol, it is essential to invest the time and effort to make this part of your program as objective and efficient as possible. The time, effort, and money that it takes to send auditors into a facility will be optimized if you are able to turn the results into demonstrable, sustainable improvements.

Below are the fundamental elements for an effective audit protocol:

Repeat the standards from your code

- State the code section (e.g., health and safety)

- Create an identifier for each standard (e.g., Health and Safety Protocol #1 could be "H&S 1")

- Repeat the wording for each standard in the code section verbatim (e.g., "The facility provides appropriate controls to mitigate health and safety risks in the workplace")

Provide clear procedures for assessing conformance to the standards

- Identify the documents that must be reviewed (e.g., worker injury and illness records, results from exposure assessments, and specifications for personal protective equipment)

- Specify the interviews that must be conducted (e.g., facility health and safety manager, and employees who work with hazardous chemicals)

- List the required physical inspections (e.g., manufacturing equipment that uses hazardous chemicals)

Establish a method to score compliance

It is a common misconception that compliance is a binary function. In practice, you might find a very minor issue that should be noted (like one fire extinguisher at the wrong height on the wall) vs. a widespread issue (like no fire extinguishers in the facility). Consider establishing a scoring system to rate the level of compliance (or the prevalence of the violation). For example, if the violation is widespread, the score might be one out of five. If the violation is observed once or twice it might be considered "isolated" and scored three out of five. If there are no violations, you would assign a score of five.

Severity is another important aspect to grading your compliance findings. Most programs will categorize findings as "minor, major, or severe" (Apple referred to severe violations as "zero-tolerance").

For example, if your audit team discovers underage workers on the manufacturing line or conditions that create an imminent threat to health or safety, such as blocked fire exits or heavy equipment with no safety guards, you might consider these severe violations. Typically, any severe violations require an immediate response and correction by the supplier. Similarly, major violations require faster resolution than minor non-conformances. The EICC is a good reference for severity categories, but I would recommend conducting a full review of these categories and making adjustments for your program.

Establish a method to score management systems

The difference between a good and great audit protocol is the focus on *management systems*. Compliance findings are really just a snapshot of the conditions when the auditor was in the factory and don't tell you much about how the supplier is managing risks in its factory on a sustaining basis. Again, you could set up a scoring system to evaluate the management systems for *each element of the code*. These scores can evaluate whether the supplier's management system has a policy, procedure, qualified staff, feedback systems, and a process to correct problems.

You can review the scores for compliance and management systems independently or combined. Apple multiplied the management system scores with the compliance scores to get the total score for each item. This meant that even if a supplier was in full compliance (five out of five) for an item, but had no system to manage this area of the code (one out of five), they would have received the ultimate score of five out of a possible 25 points. This may seem very tough for a company that is in full compliance, but without a strong management system, compliance will not be sustained over the long term.

Instruct auditors on recording their findings

There is an important distinction between audit findings and notes. Findings should succinctly state the issue: for example, "five of ten worker dormitory rooms found with triple bunks, which are not allowed." The notes can go into more of the subjective details about the finding: for example, "workers seem crowded in their rooms and several complained that they are unhappy with the situation." While notes providing background information can be helpful, overly detailed reports create ambiguity and unnecessary work for anyone who is trying to understand the problems and work on solutions.

Working in China

At this point there are likely a few readers thinking, "Wait. Workers live in dorms?" It is an eye-opening experience to visit factories in "low-cost economies" such as China, where it is fairly common to provide accommodation and meals for workers. Many of these workers have traveled far away from home for these jobs and do not have the resources to afford reasonable housing and/or food. Without the room and board provided by their employers, these workers would have a much more difficult time supporting themselves because most of their wages would be taken up with living expenses. Many companies provide living facilities as a benefit in addition to salary – or with a substantial subsidy. This allows the workers to save more of their wages while the company ensures that the workforce is well cared for and thus more reliable.

At Apple, I made it a point to personally conduct at least one audit every year and met with the managers of major supplier factories at least once per quarter. There is no substitute for visiting supplier facilities in person to get a sense of the culture at the factory and how things really work. One of my favorite tactics was to wait until the end of a long day of meetings or auditing and, without prior notice, tell the factory managers that I wanted to stay in the worker dormitory for the night. I did this a couple of times and, even though the factory managers found an empty room for me, I got a first-hand sense of how it feels to sleep in a worker dormitory. In addition, with the help of an interpreter, I was able to talk with some of the workers who were living on my floor to understand their perspectives and make a few friends.

The audit procedure

In addition to the audit protocol, you should outline a step-by-step audit procedure that lays out all of the steps for *how* to conduct an audit. The audit procedure should set out the instructions and the sequence starting with pre-audit preparation to following up on corrective actions after the audit. This document will help prepare both your audit team and the audited facility and make the audit far more professional and meaningful.

Below are the fundamental elements for an effective audit protocol:

Audit preparation

Successful audits start with good preparation. The audit team should research the facility and the company being audited and review any existing documentation such as prior audits, self-assessment questionnaires, local laws, culture and customs, etc. At this stage, you should also scope out how long the audit might take, develop a schedule, ensure that you have an adequate number of appropriately trained auditors, and that each has been given an assignment appropriate to their skill areas. The audit team should meet to discuss their plan and schedule, as well as when to notify the supplier. In some cases, your team may choose to conduct an unannounced audit, but in my experience it is best to give the facilities at least two weeks' notice so that they can have the staff and records in place for your team. In some cases, the facility management may either "clean up" their operation or even falsify documents before the auditors arrive on site. It is important to ensure that your audit team is trained to detect falsified documents, interview people, and inspect management systems to try to get as close as possible to the real-world conditions at the facility.

Selecting the right audit team is also essential to a good audit. Because auditing is tedious and time-consuming work, most companies will use a third-party auditing firm to perform their supplier audits. The typical auditor from these firms is young and inexperienced. Regardless of what your auditing firm may tell you about their auditors' experience, in practice I have found that it was vital that all of the contract auditors were trained and tested on our company's specific audit protocol before they were permitted to conduct audits. Also, I considered it a best practice to have one of my company's employees supervise each and every audit in person. Not only is it good to get first-hand knowledge of your suppliers' facilities, but the supplier realizes that your company is serious about the process and is likely to take it more seriously. In addition, having an employee on-site will lessen the chance of any corruption of the auditors.

Corruption in auditing (such as bribing auditors or falsifying records) is real and widespread. To avoid this, make sure that you conduct background checks on your auditors and have other checks in place to ensure that they are not taking bribes. In addition to monitoring your auditors' ethics, your team should be on the lookout for fraudulent documentation. For example, your auditors should compare the supplier's work records against your company's production schedule and you might discover that the workers have been working "off the books."

The opening meeting

When your team first arrives on the site, they should start off the audit with an opening meeting with the facility management. This meeting is where your team will lay out why you are there, run through the process and schedule, and, most importantly, state that *the audit is an important part of your overall business partnership*. You should seek to lessen the inherent tension from an audit experience by letting the facility managers know that the results are opportunities for improvement, and not a litmus test for continuation of business (there are exceptions to this if the findings are truly egregious, but even in these cases, suppliers typically have an opportunity to correct problems). The meeting should include time for the supplier's facility management to present their programs and results as well as orient your team to the site. Tactical details should be discussed, such as where the audit team will set up, the audit schedule, and who will be available to help the team in each area.

Conducting the audit

The audit usually begins with a facility tour to get a good orientation of the facility and to understand where any safety hazards may exist. After the tour, the team should split into groups. For example, your labor auditors might go with the HR manager to review records and select workers for interviews, while your health and safety auditor might meet with the facility safety team for a detailed inspection.

Interviewing workers is a particularly sensitive task. Your team should prepare in advance to make sure that you have the appropriate language skills and/or translators on the team. Sort through the records yourself to make sure that the workers are selected at random. To the extent feasible, conduct the interviews anonymously in a confidential space without the facility management present. Since it is next to impossible to completely protect worker identities, make sure that the facility managers and the workers know that retaliation will not be permitted and is a zero-tolerance violation that could result in loss of business. You might also hand out a hotline number for the workers to use if they feel that they were pressured as a result of the interview. Make sure that you select contract workers for interviews as well as workers directly employed by the supplier. Contract workers are often treated differently and may be obligated by an illegal bond.

The audit team should meet to compare notes, findings, and scores at least daily during the audit. Many times, the overall picture does not become clear until the team has had a chance to discuss the issues as a group. These meetings can also lead the team down different pathways based on the results. Make sure that these meetings are confidential, and include only the audit team (you may want to conduct these at your hotel) and *do not share any findings with facility management before the end of the audit*. Your hosts will likely pressure you for this information, but it is best to keep the information confidential until the team has a chance to discuss and verify the results. In my experience, initial findings are not always accurate and should be verified by further digging. The exception to this rule is if your team turns up a situation that might be an imminent threat to health, safety, or the environment or could result in serious criminal liability. Your audit

procedure should define these circumstances and the appropriate procedures for immediate notification.

Closing meeting

After the audit, you should hold a closing meeting with the supplier's top management to present the preliminary findings of the audit.
Emphasize the preliminary nature of the findings and give the facility
managers time on the agenda to react and possibly refute any audit
findings that they consider inaccurate. In cases where findings are in
dispute, ask the facility managers to follow up after the meeting with
the data or information needed to resolve the finding. In my experience, these meetings can be tense, especially if the findings are serious. Make sure that you are well prepared for the closing meeting (this
might mean dropping out of other auditing tasks) and ensure that you
fully understand the basis for the most serious audit findings.

Your lead auditor should open and close the session with the context of the audit (e.g., a partnership based on mutual expectation of
continuous improvement) and process for follow-up on the findings
(e.g., all findings should be resolved within 90 days – with shorter
time periods for more severe findings). Next, outline the schedule for
the draft audit report, how long the supplier's team will have to comment on the report, the final audit report, the corrective action plan,
remediation report, and a follow-up visit. Have each of your auditors
walk though their findings in succession and allow the facility management time to respond to your presentation. Make sure that the most
senior manager from the facility is present for this meeting and physically signs a document saying that he or she received the results.

After the audit

A couple of weeks after the audit send the draft report to your supplier
for review. After the supplier's review, issue a final audit report and
request that your supplier provide a corrective action plan to address
any deficiencies in the report. By asking the supplier to propose the
corrective actions, it forces them to assume responsibility for the
results of the audit.

The audit cycle is closed when all corrective actions are resolved. You should set out a definitive schedule for when the supplier will submit documentation of closure for each finding (as discussed above, you should set a shorter schedule for closure of the more serious findings). After closure, schedule a quick verification visit to check whether the corrective actions were completed appropriately. In practice, verification visits can generate even more findings, which can start a new cycle. Depending on the results of the verification audit, you can issue an audit "closure letter" to document the results of the audit. The closure letter should also set out the expectations for future audits: for example, if the audit and closure process did not go well, you would likely want to audit the facility again within a year, otherwise you might look at a longer schedule.

With the amount of detail and effort that can go into an auditing program, it is not surprising that these programs make up the bulk of supplier responsibility programs. As stated above, however, these programs are necessary but not sufficient. By limiting your actions to compliance, you run the risk of permanently assuming responsibility for your supplier's compliance performance. In other words, the cycle of auditing and corrective actions takes over from the supplier's own responsibility to responsibly manage their facilities. To avoid this, your program should apply the elements below in balance with auditing and compliance actions.

Program element 2: Business integration

A theme repeated throughout this book is that corporate responsibility is often considered an add-on to the primary functions of the business. If your supplier responsibility program is relegated to a niche that is considered "odd" or less important than product quality, engineering, or other business issues, your suppliers will pick up on this and your program will likely suffer. It is critical that the most senior procurement managers within your company visibly support the supplier responsibility program.

The most effective way to ensure that you have support for supplier responsibility issues is to integrate the program into the standard business processes between your company and the suppliers. There are three important business processes to integrate with your supplier responsibility program: supplier selection; supplier business reviews; and supplier termination.

Supplier selection

The most opportune time to inject responsibility into the customer/ supplier relationship is during the "courtship phase." When a supplier is working to get your company's business, it will be particularly attuned to customer needs. The trick is to get responsibility elements integrated into this process. Often, there is a high degree of confidentiality surrounding these negotiations because they can be material to one or both companies. It is not realistic to expect that the corporate responsibility department will be included in all of these negotiations, but below are a few tips for integrating your program into the supplier selection process.

An effective and efficient strategy is to train the people involved in supplier selection to ensure that responsibility criteria are part of the initial screening process. For example, you might ask for supplier responsibility terms to be included in the standard request for proposal document and contract template for new supplier agreements. Another good step is to ask the prospective supplier to complete a self-assessment questionnaire. Your team can assess the results and/ or train the people in your purchasing department to spot any concerns and alert you. Depending on the situation, you may also want to conduct a baseline audit of new suppliers before your company starts doing business with them. Some companies do this as a standard procedure; others use pre-engagement audits only when there are significant risk factors.

Supplier business reviews

A standard business practice in most customer–supplier relationships is the supplier business review (SBR). Often these are conducted quarterly and referred to as quarterly business reviews (QBR). In my experience, the SBR can be *far more effective than auditing* at driving performance improvements. The reason why SBR meetings are an effective process to evaluate supplier performance across a broad range of issues is that *the outcome has a direct impact on future business awards.*

SBR meetings consist of a series of "grades" from each of the major functions that interact with the supplier. For example, the engineering group may rate their experiences over the last quarter on a scale of 1 to 100, as will the quality department and procurement, etc. As the grades are discussed, the specific issues behind the grades are discussed to identify deficiencies and drive improvements. Typically, at the beginning and/or the end of the meeting, the grades are aggregated into an overall assessment of the supplier's performance. In most instances, these meetings will involve very senior people from both companies so that each group gets a full understanding of the issues in the business relationship.

Including supplier responsibility as one of the graded areas that are discussed at the SBR meeting is so effective because it makes it clear to the supplier that responsibility is a fundamental part of the business relationship. This process also exposes senior managers from both companies to responsibility issues, which raises awareness and increases the likelihood of action.

If you are able to inject responsibility issues into the SBR process, there are a couple of tips that will make this opportunity far more successful. First, establish a series of key performance indicators (KPIs) (see Chapter 4 under the section "Measure the right things"). I have found that it is most effective to ask the suppliers to submit the KPI data outside of the audit process discussed above. This gives the supplier a chance to show what it is doing right, rather than focusing on the problems discovered in an audit. In addition, if you collect

KPIs on a quarterly or semiannual basis, it results in a more timely representation of actual facility conditions.

Once you have defined your KPIs and have data coming in from your suppliers on a regular basis, use this information as an objective part of your grades in the SBR meetings. The KPIs should be one part of your overall grade along with audit results and other topics. I like to hold an SBR preparation session with each supplier to dive into the details around an established agenda of issues. This session will give you the information you need to formulate a grade for the formal SBR meeting.

There are two important aspects of using KPIs in an SBR process:

- **Selecting the right KPIs**. Select indicators that give you the best sense of the overall health of the supplier's facilities without measuring every item in the code. Items such as the numbers of workers exceeding 60hrs/week, the number of workers working longer than seven days without a day of rest, the worker injury and illness rates, and environmental emissions, incidents or violations are all good examples of metrics to get a handle on the inner workings of supplier facilities

- **Trends and comparisons**. The most effective uses of KPI information are trending and comparisons. Trending means that you simply track the supplier's performance on each indicator over time (e.g., quarter-over-quarter performance). Comparisons look at how the supplier is performing compared with competing suppliers. In my experience, comparisons with competing suppliers are perhaps one of the most effective means to drive improvements. It is advisable to keep the names of other suppliers in the category confidential when you are making comparisons

Supplier termination

It is rare that a supplier is terminated solely because of corporate responsibility issues. In fact, it is most likely that a combination of factors that have built up over time lead to the decision to switch suppliers. Nonetheless, it is important that responsibility criteria are represented in this process. Again, inclusion in the termination process makes it abundantly clear to your suppliers that responsibility is a critical part of the business relationship. In some cases, termination of the supplier can become a major motivational force if other suppliers recognize that their performance on responsibility issues can be pivotal to their entire business relationship – in other words, actions speak louder than words.

A common misperception is that suppliers are automatically terminated for serious code of conduct violations. In my experience, this is the rare exception and frankly not a wise choice. When a serious violation is discovered – child labor for example – it is far preferable if the supplier fixes the problem, rather than simply ending the business relationship. Ending the relationship leaves the problem unresolved and may even make conditions worse. By working with suppliers to resolve serious issues, the buying company can use their economic influence to sustainably improve conditions. For example, if underage workers are found, the resolution should include returning children to their parents and ensuring that there are management systems in place to prevent more children from being hired. This is a much better outcome than pulling away. Leaving these problems unaddressed means that the supplier may do nothing to resolve the underlying issues and they may worsen.

Program element 3: Capacity building

The Chinese proverb: "give a man a fish and you feed him for a day, teach a man to fish and you feed him for a lifetime" applies to this program element. Capacity building is a term borrowed from international development that is defined as: "Understanding the obstacles that inhibit people from realizing their developmental goals while enhancing the abilities that will allow them to achieve measurable and sustainable results."

This is a very appropriate definition because the basic idea is to enable your suppliers with the essential capabilities and self-sustaining systems so that they will be able to independently comply with the code of conduct.

Some people in your company may object to this part of the program because it goes one step too far in getting involved in your suppliers' business. Companies like Nike and Gap that have been working on supplier responsibility for some time, however, understand that it is far more cost-efficient and effective to ensure that suppliers have the basic capacity to achieve compliance than fall into an endless cycle of audits and corrective actions.

For example, one capability gap found in some factories I have audited was that the management could not identify a trained professional with responsibility for environmental compliance or worker health and safety. Without a professional in charge of these important issues, it is almost guaranteed that there will be violations and perhaps serious mishaps. Another example is when the rank-and-file workers have not been trained on their basic rights or how to appropriately raise a grievance.

Start by identifying the gaps in your supplier's capabilities by reviewing audit reports (specifically the management systems element of the reports). Once you understand the gaps, your capability-building plan could be to persuade your supplier to hire a skilled professional in a critical position or to provide additional training for their employees. For example, an often-used capability-building solution is to ensure that assembly-line supervisors are trained in the appropriate techniques to work with their subordinates in a respectful way and the workers are trained on the workplace standards as well as their rights and remedies.

Program element 4: Communications and stakeholder relations

The aim of this program element is to explain your overall supplier responsibility program to the outside world and, in turn, understand their perspectives. It can be very difficult to communicate about supplier responsibility matters. Not only are customer–supplier relationships typically confidential, but the types of issue discovered in supplier audits can also be very sensitive. There are often non-disclosure agreements (NDAs) between customers and suppliers that make the situation even more complex. Nonetheless, it is critical for your program's credibility to communicate results external to your company. Without this kind of communication, the default assumption will be that your company has done nothing to manage supplier responsibility, heightening the risk of becoming a target for activist campaigns. In addition, transparency has other benefits: the act of pulling all the data together for a public report becomes a program review and quality check. In every report I have ever been involved with, we found significant areas for improvement by looking at the patterns in the data.

To report supplier responsibility publicly and still protect confidentiality, most companies will aggregate their findings into categories. For example, companies may report the number or percentage of violations in each category of their code. Reporting data this way provides significant information without revealing supplier names or associating violations with individual suppliers. A few leading companies, such as Hewlett-Packard, have taken the step of identifying all of their major suppliers. But even these leaders have not published which of their suppliers have committed specific code violations.

If reporting is a one-way form of communication about your programs, then stakeholder engagement is a two-way dialogue. As mentioned earlier in this book, many companies get a wake-up call on supplier responsibility issues from activist groups that use the press or shareholder proxies to shine a light on poor practices. I have called this the "2x4 effect" as a metaphor for the "whack" that is felt when your company is the target of a public relations campaign. Whether you have suffered through a "whack" or you are simply building a solid program, it is wise to build relationships with the stakeholders who are following your company and who are most likely to comment on your performance.

There are two old phrases that I have used over my career that fit this area. First, "Keep your friends close and your enemies closer," which was attributed to Sun-Tzu, a Chinese general and military strategist. The other phrase that works well for stakeholder engagement is a quote from President Lyndon Johnson, who commented about his political opponents, "It is preferable to have them inside the tent and pissing out than outside the tent and pissing in."

Whichever phrase you identify with, the meaning is the same: get to know the people or groups who are most likely to be critical about your program. Directly reaching out and taking the time to sit down (face-to-face is best) to talk about your progress and their concerns will pay off in several ways.

No, you will not eliminate their criticism. If you enter these kinds of discussion with that as a goal, you may become bitter and frustrated. By definition, activist groups conduct "name and shame" public relations campaigns to raise awareness. Instead of trying to eliminate criticism, your expectation should be to build understanding, common ground, open communication, and good relationships. By opening communications and building relationships, you will be better able to anticipate activist concerns and take steps to address them before they end up in newspaper headlines. You will also be able to communicate all of the actions you are taking to improve performance and, hopefully, build some credibility and trust. In addition, if you enter these discussions with an open mind, you may learn something that will improve your program. Finally, with stakeholder relationships based on credibility and trust, it is less likely that your critics will attack you in the press, or they will at least give you a heads up before they do.

Communications and stakeholder relations will be discussed in detail in Chapter 8. For the specific area of supplier responsibility, though, it is essential for you to identify the stakeholders who are tracking these issues for your company and proactively engage them in the implementation and planning for your program.

Conclusion

After reading the chapter, you should have a high-level understanding of how to construct and operate a supplier responsibility program. As with the rest of the book, the information is deliberately designed to give you an overview of the tips and tricks that work *in practice*. But, like the disclaimer on a diet plan, "your results may vary." There are infinite variables that will affect how you apply this information, with the big ones being your company's culture and your position in the company. Keeping this situational element in mind, however, you should be able to pick and choose from these concepts to compete for a position in supplier responsibility and instantly add value.

One broader philosophical thought to close this chapter: With globalization opening up access to cheap labor and lax environmental regulations around the world, many pundits believe that there is now a "race to the bottom." The classic paradigm is that business will seek the most expedient way to make a profit and, in doing so, likely cause harm to the environment and/or abuse workers.

The field of supplier responsibility turns this concept on its head. Instead of a race to the bottom, we are seeing a race to the top. Companies are increasingly judged and compete on the basis of social and environmental responsibility. Government is not driving this, since most regulations don't even apply when business is done in other countries. These improvements are being accomplished because activists and media have been successful in shining a light on poor supply chain practices and linking these to major brands. Now, companies around the world are focused on the social and environmental practices of their suppliers as never before and are driving improvements throughout the global supply chain.

When I ran the Apple supplier responsibility program the company was criticized for trying to impose Western values on other cultures. You may encounter similar criticisms about your program being "political correctness run amok." This is complete hogwash. The Apple code, and every other code of conduct that I have ever reviewed, is based on well-established international standards and guidelines. These are not capricious musings about utopian business operations.

Codes of conduct and their implementation are one of the best forms of corporate responsibility. Holding an entire supply chain accountable to a set of basic expectations can, and does, have significant and measurable positive benefits to people and our planet.

Supplier responsibility programs are ultimately about creating a supply chain that treats people with respect and dignity and values the environment. Without question, the ability to drive tangible improvements for workers and the environment through supplier responsibility programs has been a highlight of my career. With the very real incentive of billions of dollars of business on the line, the motto for these programs could be: "our dollars, our values."

8

Communicate! Part 1
Talking about corporate responsibility

This chapter outlines specific tips for building your skills in spoken communications about corporate responsibility both internal and external to your company.

Action speaks louder than words but not nearly as often (Mark Twain).

At the core of the corporate treehugger's skill set is the ability to communicate well. Whether you are in the environmental department, supplier responsibility, or corporate responsibility, the ability to communicate is the single most important (and perhaps the most overlooked) skill. It comes in several forms: public speaking (conferences and stakeholder events); writing (the corporate responsibility report, blogs, and even tweets); and influencing others within your company (aligning on a common strategy).

The ability to communicate well verbally can be a touchy subject because it is closely integrated with your personality. Some of us are extroverts and thrive by talking (though simply loving to talk does not a good communicator make). Others are introverts and prefer their own company to interacting with others (conversely, introverts can be excellent communicators because they think through what they will say before they say it). Whether you love to talk or are happier with your own "inner voice," verbal communication skills can be learned, and with practice you can improve your abilities.

Early in my career, I was a basket case in this area. I don't think I realized how bad I was until I started to work at EPA and lost a whole night's sleep before I had to give a presentation. During the presentation, which was to a group of senior people, I had all the symptoms: red face, talking fast, forgetting what I wanted to say . . . most of us know the drill.

All of the people I admired as leaders were also powerful speakers. When you think about it, the word leader means that people follow you. By definition, people who are great leaders create compelling messages that cause others to follow. If I ever hoped to achieve my goals, I knew that I had to overcome my fear and find ways to become an effective communicator.

Beyond my own career ambitions, there was another realization that motivated me to become a better verbal communicator. I have a technical background and, early in my career, I worked mainly with other technical people who had similar expertise. While we could communicate with each other about our work, it was unlikely anyone else would understand what we were even talking about. This deficit was profoundly limiting. Even if I became the best within my technical field, my insights and innovations would be locked within a small circle of collaborators. Like a PC without the Internet, knowledge that is not effectively disseminated is unusable by others and thus limited in its power to change the world. By honing and mastering your communications skills you can turn knowledge into power by creating compelling, understandable, memorable, and even inspiring messages. Becoming an excellent communicator is the single most important accelerant to your career path both within and beyond the field of corporate responsibility.

In the field of corporate responsibility, the ability to communicate well is even more important than in other career pursuits. Unlike other careers that are focused on a particular discipline, corporate responsibility requires that you understand and can easily represent almost all company functions. In this respect, corporate responsibility is similar to a career in press relations or marketing. For example, my training and early career was in environmental management. This is a field based on chemistry, toxicology, exposure modeling, risk assessment, laws, and regulations – all very detailed and technical information that takes years to master. Even though I had made it a personal goal to improve my communications skills, success in this field hinged mainly on my technical ability within a fairly narrow area. As I became more steeped in the jargon of this field, my ability to communicate outside of the field diminished.

Now, as my career has evolved into the field of corporate responsibility, the crucial skills are no longer technical, and my success is directly attributable to honing my communications skills. Corporate responsibility managers must aggregate, and essentially create, the "CR story," which spans almost all company functions from environment to ethics. AMD refers to corporate responsibility as "the sum total of our behaviors." What this means for me is that I must understand AMD's culture, history, and a broad range of corporate functions and be able to speak to the press, social investors, NGO activists, and others in a manner that is at once candid, accurate, and supportive of AMD's reputation.

This chapter will explore the specific application of verbal communication skills to the field of corporate responsibility within a company and, because you will likely be called on to be a company spokesperson, it will also cover speaking to external audiences. The foundation for mastering verbal communication skills is an entire field in and of itself. The ability to speak in front of large groups of people is the first thing most people think of in this area. While it is not the only facet of verbal communication skill, public speaking gets the most attention because statistically it is one of humankind's top fears. Jerry Seinfeld got a big laugh when he joked about a survey that found that the fear of public speaking ranks higher in most people's minds than the fear of death. "In other words," he deadpanned, "at a funeral, the average person would rather be in the casket than giving the eulogy."

In my personal journey to improve this skill, I have greatly benefitted from several company-provided training courses on how best to structure and deliver oral remarks. As I improved, I even became an instructor for the presentation skills course within Intel. I highly recommend this kind of training and, even more importantly, practice! While I practiced in primarily "on-the-job" settings, I would recommend Toastmasters International for both training and practice.[62] The first paragraph from the Toastmasters website embodies the importance of mastering verbal communication skills:

> Confidence. The ability to communicate, persuade and lead. The skill to tell one's story, shape better tomorrows and point others in the same direction. These are the attributes of leaders, and not all leaders are born with talent. They learn it, and so can you.

The Toastmasters organization is dedicated to improving presentation skills and has professional chapters throughout the world. Their meetings provide a safe venue to practice and polish your skills with a supportive audience that has no input on your annual performance review!

Corporate responsibility communications inside your company

The first thing to understand is that many, if not most, of the people with whom you will communicate inside of your company have very limited understanding of corporate responsibility. Internal stakeholders are consumed with their own job functions and, unless they are personally interested in the topic, are unlikely to know much about the theory or practice of corporate responsibility. This is natural because, unlike traditional business functions like sales, manufacturing, finance, legal, marketing, etc., corporate responsibility is a relatively new addition to the business scene.

The elevator speech

In just about all of my internal communications, I start with a concise "elevator speech" that introduces my function. An elevator speech is a very brief description of your role that takes as much time as a ride on an elevator: no more than 60 seconds. A corporate responsibility elevator speech might go something like this:

> Corporate responsibility is defined as balancing the "triple bottom line" which is essentially "people, planet, and profit." Our team monitors, manages, and communicates with the public about the company's behaviors in several areas including environment, ethics, labor issues, supply chain responsibility, and others that together make up the reputation of our company. We help attract social investment to the company as well as build and protect the company's reputation with customers, employees, and other stakeholders.

Write down your elevator speech and practice it out loud a few times until you have it mastered. This does not mean you have to memorize it, but you should be comfortable enough to relay the essence of corporate responsibility to anyone instantly and briefly.

The ROI trap

Notice that the elevator speech concludes with a statement about the value offered by corporate responsibility. As discussed earlier in this book, the corporate responsibility function is frequently challenged regarding the value it returns to the company. I call this the "return on investment (ROI) trap." Corporate responsibility managers will run into the ROI trap often and you should be prepared to respond to this challenge – it can go something like this:

> I don't get it. Tell me again what you do?
>
> How does that add value to the business?
>
> So, is this really just PR?

Depending on the situation, it can be hard to respond to these challenges in a civil and respectful manner. Loaded within the challenge is the notion that your function is somewhat superfluous or, at a minimum, less important than that of the person who issues the challenge. Before exploring how to respond to the ROI challenge, it is important to understand some background. All business functions must prove their ROI. Like the immune system in your body, a well-run company will tend to reject any new additions that do not demonstrably add value. The more "mainstream" functions do not face as much of this scrutiny because they have been in existence longer and thus their intrinsic value is well accepted. Corporate marketing, for example, is accepted as an essential element for businesses of all types. While the marketing department typically spends a significant amount of company money for somewhat intangible results, such as building the brand, it is accepted as a needed element to drive sales and boost the company's reputation.

Like marketing, the effectiveness of corporate responsibility is evaluated on less quantifiable measures of value such as brand reputation. The difficult-to-measure benefits, combined with its relative newness, make corporate responsibility a target of the corporate "immune system" and leave its practitioners vulnerable to pointed questions about the ROI of their jobs. I call these questions the ROI trap because, by vigorously defending your function's right to exist (a natural reaction), you can easily reinforce the view that corporate responsibility is superfluous. Becoming entangled in a dispute over intangible benefits is unlikely to convince your colleague of anything other than that you are overzealous and perhaps a bit defensive.

A better course is to acknowledge that the benefits of corporate responsibility are, like those of marketing, hard to measure. Time to practice your corporate Jujutsu: You should "size up" your questioner to assess whether they are genuinely interested in exploring the qualitative benefits of corporate responsibility or are dug in to their anti-CR stance. If it is the latter, look for a graceful or even self-deprecating way to leave the discussion. It is certain you will encounter many entrenched CR doubters in your corporate career and, in most cases, it is better to avoid any attempt at evangelizing your role to them.

Without acknowledging their views, you can avoid "rising to the bait" by turning the conversation back to their role in the company (likely their favorite topic) and seeking areas of mutual interest. In other situations, where your colleagues are truly seeking to understand the value of your function, there are a number of ways to engage in the ROI dialogue:

- **Employees.** Chapter 11 of this book covers the linkage between employee engagement and corporate responsibility in depth. For your internal discussions on ROI, it can be useful to memorize a few data points on this linkage and its importance to the business. For example, the Gallup organization dubbed employee engagement "a leading indicator of financial performance" and backed it up with research showing that "engaged organizations have 3.9 times the earnings per share (EPS) growth rate compared to organizations with lower engagement in the same industry."[63] The 2010 Hewitt Associates study "Engaging Employees through CSR" showed a strong correlation between engagement and people who believe their organization is socially and environmentally responsible.[64] As you will find in Chapter 11, there is an increasing body of evidence that employees are motivated and more productive when they perceive their company as being devoted to responsibility

- **Investors.** The large and growing assets in socially managed investment funds are an important factor in the ROI calculation for corporate responsibility. Having a strong corporate responsibility program is, of course, essential to attracting socially managed investment dollars. Again, memorizing some statistics is useful. Borrowing from Chapter 10: "The Social Investment Forum estimates that, as of 2010, the total professionally managed assets following socially responsible investment (SRI) strategies amounts to $3.07 trillion and is growing at a faster pace than conventional investment assets."[65] In addition, because many of the large institutional retirement funds are moving into social investing, their buy-and-hold investment strategies can reduce the volatility of a company's stock

- **Brand and reputation.** Most business professionals accept that the company's brand is an important asset and would not question the resources applied to maintain and improve the brand image. In Chapter 13 I make the point that, for many companies, their brand is the largest intangible asset on their balance sheet. Apple led the 2011 valuation (from Millward Brown) with a brand valued at $153 billion.[66] Couple these eye-popping brand valuations with the finding from the Reputation Institute that over 40% of a company's reputation stems from its corporate responsibility programs, and the connection between CR and company value becomes obvious. Even without these statistics, most people intrinsically understand that a company's reputation is central to its success with customers, suppliers, employees, and stakeholders. You should point out that, at its essence, corporate responsibility is an increasingly important driver of corporate reputation

- **Ethics and activists.** Today's business managers are acutely aware of the increased scrutiny on corporate behavior sparked by numerous high-profile business ethics scandals. In describing the value of corporate responsibility, you should make it clear that it is tied to the ethical performance of the company. The CR department works directly with activists to discuss their concerns with the company's ethical policies and performance. In essence, you are working on the front line to head off PR disasters that, based on history, can take down an entire company

The subject of ROI is at the heart of many internal communications about corporate responsibility. Being well prepared for these discussions, and avoiding the traps and pitfalls of being drawn into dogmatic debates, are important skills for the corporate treehugger. The tips presented here are necessarily generic. It is always best to personalize your communications to your company's programs. The key points to remember in these discussions are:

- Be prepared with your elevator speech that covers value to the company

- Avoid being drawn into dogmatic debates that serve only to fray relationships

- Memorize a few of the basic facts that demonstrate the value of corporate responsibility: employee engagement, brand reputation, and social investment

The case against CSR

In 2010, University of Michigan business professor, Dr. Aneel Karnani published an op-ed in the *Wall Street Journal* titled "The Case Against Corporate Social Responsibility."[67] Dr. Karnani's article seemed almost deliberately provocative and, since it was published, he has reaped considerable notoriety as he travels to events to defend his views. I observed a debate between Dr. Karnani with Gerald Sullivan (President of the Vice Fund which invests in tobacco, alcohol, and weapons makers) vs. Paul Herman (CEO of the HIP investor) and Dr. Vinay Nair (Adjunct Associate Professor Finance and Economics, Columbia Business School) in 2011. Had this debate been a boxing match, it would have been stopped in the early rounds. Herman and Nair presented reams of compelling data supporting linkage between corporate responsibility and business value. Karnani and Sullivan mostly mewed about their opinions and at one point openly disagreed about the role of government. (Karnani's view is that government regulation, not corporate responsibility, is a better system. Sullivan, like all good capitalists, is not a big fan of government regulation.)

When Karnani's article was first in print, I published a counterpoint (along with hundreds of others) in an article titled "The Case Against the Case Against CSR," which covered three main reasons why corporate responsibility adds value to the business – below is an excerpt:

More companies are winning with CSR
Companies are looking at the megatrends in the world and asking themselves: "how can we apply our core competencies

→

to win in the future?" It so happens that many of today's trends point to sustainability issues – resource scarcity, poverty, pollution, etc. While a litany of doom for some, these issues can also look like opportunities for a wise business manager. For example, General Electric CEO Jeff Immelt has positioned GE for success in a resource-constrained world with the "Ecomagination™" line – topping more than $18 billion in revenues in 2009 and a growing profit center.

Smart companies take the long view

Dr. Karnani warns that CSR may be dangerous because, by doing the right thing voluntarily, companies may obscure the need for government regulation. Also, why would companies take costly actions beyond what is required by regulations? Leaving aside moral issues, the answer lies in taking a longer view. Sure it may require additional investment to responsibly manage a business, but when left unchecked, poor conditions can go awry costing many thousands of times more. For example, following Karnani's logic, companies operating in countries with poor or nonexistent laws for toxic waste disposal should just dump their toxic waste out the back door or into the local river since it is cheaper and legal to do so. Even if this was a moral choice (it is not), many of these companies have learned what can happen if they wantonly pollute the environment. The US Superfund law made companies responsible for cleaning up their mess regardless of whether dumping was legal at the time and often stuck them with the bill to clean up waste dumped by others as well. Companies learned an expensive lesson: it is cheaper to manage their waste responsibly than get stuck with a huge bill and a bad reputation.

CSR impacts brand value and investment

Often listed as the largest intangible asset on the balance sheet, brand reputation can make or break a business. Rather than struggling with definitions and rationales for CSR, most companies intrinsically understand the business rationale to act responsibly and, if possible, lend a hand to make things better. They know that their reputation and their brand depend on it.

While it may not have been his intent, Dr. Karnani's provocative opinion may have done more to promote CSR than to slow it. Having stirred up legions of current and future business leaders, he has added momentum to the CSR movement for years to come.

Executive communications: Decision briefings

It is almost certain that your career in corporate responsibility will involve communicating with the leaders of your company. As highlighted above, corporate responsibility is an integral part of the company's reputation and the company's leaders will, at a minimum, need to be informed, and likely will want to shape the strategy. Unlike a public speech, executive communications need to be pared down to the basic facts, extremely candid, concise, and focused on whatever decision needs to be made.

This book covers two basic types of executive communication: operations reviews (general update briefings with no decisions needed); and decision meetings (where you are asking for approval of a change that has strategic or resource implications). Operations reviews are covered in Chapter 4 so this section will focus on decision meetings.

Build your support

The first thing to understand about executive decision meetings is that most of the real work happens before the meeting. In my experience, by the time the stakeholders sit down in a conference room to discuss whatever decision is to be made, the key decision-makers are aware of the issue and its implications and are predisposed to approve. This does not mean that debate will not occur in the meeting. In a well-run company, executive meetings are often debates where the full implications of a decision are weighed by viewing it from multiple perspectives.

The advice here is to make the rounds with the major stakeholders before the decision meeting to understand their views on the topic. This means identifying the stakeholders who have the most authority, are most affected, or are most likely to disagree, and discussing the issue with them in advance of the decision meeting. In these meetings, you should avoid "selling" your position but adopt a listening posture. Lay out the issue, the options for decision, your thoughts on the right path and why, and ask open-ended questions about their views. The goal for these meetings is to discover stakeholders' views on the decision and, to the extent feasible, incorporate their input into the final recommendation. In some cases, if you cannot get time with the executives involved, consider working with their top deputies to gauge their boss's insights. If you find that your "ask" (the item you are seeking approval for) is unlikely to be approved, it is best to save everyone's time (and your credibility) and cancel the decision meeting. Cancelling the meeting does not necessarily mean defeat; it can mean that you need to assemble more data to make your case and/or line up additional support.

Know your facts

In Chapter 4 I recounted my preparations for operational reviews with Tim Cook (the current CEO of Apple) as equivalent to studying for a college final exam. Being well prepared is even more important for decision meetings. A good manager will ask you extensive and detailed questions about every aspect and implication of the decision you are bringing forward. Your role is to be the expert – their role is to make the decision. To do your role well, you will need to be well prepared. Start by assembling all of the available facts and data about your topic. Next, look for the obvious gaps in the data and fill them. Finally, work with a colleague or confidant to test your approach by playing devil's advocate. Your colleague should try to poke holes in your recommendation and the supporting information. This process will not only show you the weak points in your case, but it will give some practice in presenting your case.

An example of this process stems from Intel's decision to take a more active role in supplier responsibility (see Chapter 6). In preparing for the executive decision meeting, my team assembled benchmarking information on other electronics firms that had been targets of activist PR campaigns, the costs of ramping up this program, the likely benefits, and the implications of not acting. We laid out a detailed action plan covering who would be involved, their roles, and the resources required. This plan had been reviewed with each of the major stakeholders in advance. While there was plenty of debate, the decision meeting affirmed our recommendation because we were well prepared and had briefed all of the major stakeholders in advance.

Begin with the end in mind

This phrase is from Steven Covey's excellent book, *The Seven Habits of Highly Effective People* and is absolutely appropriate for executive decision meetings. All executive decision meetings should begin with a statement about the purpose of the meeting. This is often articulated in the first slide titled something like:

> Why are we here?
>
> Expected outcome
>
> Purpose of today's meeting

Clearly and concisely state the reason for the meeting – e.g., decision on whether to adopt the electronic industry code of conduct and require compliance from all tier-one suppliers – at the outset of the meeting. This will set the tone for the discussion and provide unambiguous notice to the participants about the expectations for the meeting.

Present the case

After articulating the purpose for the meeting, present the background for the decision with as little material as possible. Continuing with the supplier responsibility example from above, the background could be a few slides covering:

- The definition of supplier responsibility and the code of conduct
- The growing public scrutiny of conditions in the supply chain
- Data showing the exponential rise in customer inquiries on CR topics
- Benchmarking results from similar companies

After covering the background, lay out a clear "problem statement." A problem statement is a clear and concise description of the issues(s) that need to be addressed. Our example might take this form:

> Our company's lack of a defined supplier code of conduct and oversight process is annoying our customers and creating risk of brand-damaging activism.

Next, lay out the options to address the problem. It is useful to present a "high, medium, and low" response to solving the problem statement. You should always include a "status quo" option either as your "low response" option or a freestanding fourth approach. In outlining the options, utilize a standard format for each. The format should describe the approach, the resource requirements, the groups affected, and list the pros and cons (or cost/benefits) of the option.

In discussing the supplier responsibility example with my Intel colleague, Brad Bennett, he pointed out a few additional points that made this decision meeting a success. Brad was in my group at Intel and I had assigned him to lead this briefing. One particularly effective tactic was ensuring that a few of the key stakeholders were prepared to speak up at the meeting in favor of the recommended approach. In this case, Brad and I had worked with Intel's procurement director as well as the director of corporate responsibility. During the meeting, both of these people supported the problem statement as well as the recommended solution. Brad also reminded me that he was nervous in the meeting, which for him meant that he slowed down his presentation. I recall that this was particularly effective because it gave the participants time to think through the content and consider the decision.

As a result of all of our preparation, the decision meeting took less than the allotted time. Intel became a founding partner in the electronic industry code of conduct and created a comprehensive supplier responsibility oversight program.

Backup plan

While the example above went well, I have seen other decision meetings that did not result in a consensus. As mentioned above, it is not

unusual for executives to openly debate the pros and cons of decisions; in fact, many would argue that this debate is a necessity for
good decision-making. If you find yourself in the middle of such a
debate, you should have a backup plan. A few tips:

- **Avoid becoming the sole focus of the debate.** If there is a genuine disagreement about the merits of the decision, you should
  neither abandon your proposal nor defend it to the death. The
  middle ground is to rely on the facts that support your proposal,
  while acknowledging legitimate weaknesses in the case and
  agreeing to take actions to follow up with additional research

- **Try a pilot program.** When there is significant uncertainty or
  trepidation about the recommended option, you could propose
  a smaller-scale pilot project to test out the assumptions and vali-
  date the cost–benefit model

- **Try another approach.** In some cases an executive will want to
  modify one or more of the elements in your proposed approach.
  This is a good thing because it gives that executive a sense of
  ownership in the final outcome. To the extent feasible and appropriate, you should be prepared to adapt your approach to accom-
  modate the input received in the decision meeting

- **Accept no for an answer.** Wouldn't life be grand if we always
  got what we asked for? There have been plenty of times when
  the consensus of the decision meeting is to do nothing new –
  i.e., the status quo option – or to adopt an approach other than
  the recommended option. There is nothing wrong with this outcome and it can even become a benefit if you manage it correctly.
  Remind everyone that it is important for the health of the company and the program to continuously scan the horizon for new
  issues and that through this analysis and discussion the group
  has been effective and made a good decision for the company.
  This way you and your team will get the credit deserved for all
  the work and preparation even though your ultimate recommendation was not adopted

Like many of the lessons in this book, the tips for executive decision meetings are applicable beyond the practice of corporate responsibility. While this information is consistent for many functions, when working in corporate responsibility, you have to address the added skepticism that creeps into internal discussions about the topic. As discussed earlier in this chapter, you can counter this skepticism with a strong ROI case as well as tying your proposed action to the company's culture, employee engagement, brand, investors, stakeholders, and overall reputation.

External communications

Transparency is one of the core tenets in the practice of corporate responsibility. In practice, transparency means that companies should publicly disclose reams of information about their environmental, social, and governance practices. As the field has grown and more companies have developed corporate responsibility programs, the transparency imperative has resulted in a tremendous amount of CR information being published by many companies. The result is a very crowded field, with companies of all types competing to be noticed and differentiate their programs from their competition. What this means to the corporate treehugger is that you must become very good at getting your company's message out in a way that gets noticed. In the next chapter we will cover how to tell your company's CR story in written form. Below are some tips for speaking in public forums on corporate responsibility.

Develop the message

One of the themes of this book is the enormous breadth of the issues encompassed by corporate responsibility. In Chapter 3, we covered techniques to distinguish the material (priority) CR issues for your company. As you assemble your company's public CR communications strategy, start by examining your company's material CR issues.

If your company has developed a leadership position on one or more of these material issues, then they are obvious candidates for your external CR communications.

Select one or two topics where your company has a unique or leading position to use as your messaging platform. For example, Campbell Soup Company has adopted the theme of "nourishment" as a differentiator in CR messaging. Its CR website outlines ways that Campbell's will nourish its customers, its neighbors, its employees, and the planet. The VP of Corporate Responsibility, Dave Stangis, often speaks at corporate responsibility conferences and the themes of his presentations are drawn from Campbell's nourishment message such as organic food, combating obesity, feeding the needy, sustainable agriculture, etc.

The point is to identify and develop one or two key messages that demonstrate the leadership qualities of your CR program. When you have identified these topics, work with your press relations department to develop a set of talking points and slides that you can adapt to various venues. If your PR department does not provide legal review, it is a wise idea to bring in your legal team to approve any messages that you plan to deliver publicly because any public comments made by a company official could become the basis of litigation over false or misleading statements.

Pick the right venue

As the corporate responsibility field has grown, there has been an explosion of conferences focused on the topic. Some of the larger companies send people and sponsorship money to just about every one of these meetings. If you are not working for one of those companies, you will likely have to carefully choose which meetings are the best showcases for your company's message.

The first criterion for selecting the right venue is the whether the theme of the meeting matches well with your key messages. While some of the larger corporate responsibility meetings, like the annual Business for Social Responsibility conference, cover enough ground to make them attractive venues for almost any CR topic, others are more narrowly focused. For example, Campbell's nourishment message would not fit well at a conference focused on corporate ethics.

Next, assess whether the meeting will be attended by influential thought leaders. The definition of who is considered influential varies by the topic(s) in your key message (e.g., there is a different set of people who are focused on corporate ethics than those who cover food and nourishment issues). Some companies assemble a list of the specific individuals that they consider influential. Others look for the broader categories of influential media, NGO groups, investors, academics, etc. Obviously, you want to focus your efforts on the "must-attend" events for your key CR messages. These are the meetings that will attract the largest density of influential people in the field.

The last item in selecting the right meetings for your message can be the most problematic: will your company be able to secure an effective speaking platform? Speaking slots can range from the featured keynote to being included on a large panel of speakers at an unattractive time. The bigger "must-attend" events can be very selective about their speaking slots. Sponsoring the event can sometimes (but not always) buy your company a speaking slot. If sponsorship is not a feasible route to attain a speaking slot, a few tactics that can help include:

- **Offer a senior executive as your speaker.** Conference organizers are always looking for top-level titles to headline their agenda and add credibility to their event. If feasible, recruit a senior-level executive from your company to speak and negotiate with the conference organizers for a keynote slot

- **Speak on a controversial or new topic.** New and controversial topics generate interest in meetings. Conference organizers are increasingly turning to techniques such as debates, un-conferences (where the participants select the topics), interactive workshops, etc. to generate interest in their events. By proposing a cutting-edge idea or speaking about a controversial topic, you may be able to secure attractive speaking slots for your company. Obviously this tactic must align with the company's legally approved CR messages

- **Moderate a panel.** If an attractive speaking spot is not available in a target meeting, you can still find a decent platform to deliver your message by offering to moderate a featured panel. As a moderator, you will not be able to go into much depth on your company's message, but you can touch on a few themes as well as establish your leadership by asking probing questions

Be a thought leader

The agendas of most corporate responsibility conferences are packed with speech after speech about company-specific CR programs. After one or two of these presentations, the program can feel overly promotional and can become boring. Conference attendees are interested in cutting-edge issues. They want to learn something new and gain knowledge that they can apply to their own work. If you want your company to stand out from the crowd and secure larger speaking venues, build your presentations around hot topics in the CR field and research the latest facts and examples about the underlying subject.

Creating thought-leadership presentations has to be balanced with getting out your company's message. For example, when I have presented on the topic of engaging employees in sustainability on behalf of AMD, I started with statistics about the value of engaged employees to the business (see Chapter 11) and then presented data on the linkage between engagement and corporate responsibility. I concluded by weaving the AMD examples in with examples from other companies on how to effectively use corporate responsibility programs to drive engagement. The final product provided the audience with a concise summary of leading research as well as useful models that they can immediately apply to their own companies. More importantly, it established AMD as a thought leader in this subject area and solidified the company's reputation among influential stakeholders.

Rinse, repeat

When I worked on Capitol Hill one of my mentors at the time used to say: "if you are not sick of your message, than you have not said it enough." While this axiom is most appropriate for politics, it's also instructive for getting your CR message out to the public. To attract notice to your company's message in the crowded CR field you will need to attend many meetings and make numerous presentations focused on the same material. The good news is that recycling your materials from venue to venue is perfectly acceptable. I usually develop between one and three presentations on my company's key CR messages and adapt them to the conference and audience. While some repetition is unavoidable, you should spend significant effort to update and tailor the content to each audience to make it as relevant and compelling as possible.

Conclusion

As discussed at the beginning of this chapter and throughout this book, communication skill is the single most important capability for the corporate treehugger. Improving my own communication skills has been a career-long personal journey that, like exercise, is hard and frankly easier to avoid, but, as I have improved, has been extremely rewarding.

This chapter focused on applying verbal communication skills to the practice of corporate responsibility, but underlying this advice is the need to perfect your personal capabilities. A monotone, poorly delivered presentation will undercut even the most compelling message. Whether you are an old hand or just starting out, everyone can improve these skills. An objective and sober analysis of what you do well and where you need to improve is the essential foundation for mastering this skill. Like any skill, improvement comes from good instruction and plenty of practice. The payoff for this hard work is compelling: the ability to speak with impact can connect, inspire, and motivate people and is a hallmark of leadership.

9
Communicate! Part 2
The corporate responsibility report and beyond

This chapter covers planning and execution of the essential forms of written communication for corporate responsibility as well as social media, audio, and video formats.

Either write something worth reading or do something worth writing (Benjamin Franklin).

Like spoken communication, writing is a critical tool in the corporate treehugger toolbox. Most people will be more comfortable with either spoken or written communication, but it is important for your success in this field to be good at both.

Types of corporate responsibility communication

In today's corporate responsibility departments, there are several common forms of written communication. Figure 5 outlines some of the major forms of written corporate responsibility communications as a function of the depth of their content and the frequency of delivery.

Figure 5 Frequency vs. depth of corporate responsibility communications

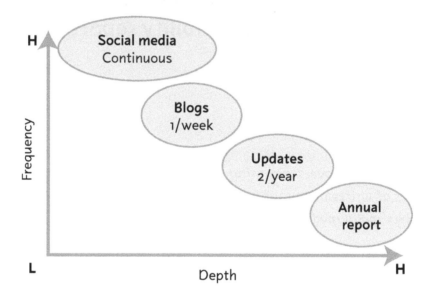

The corporate responsibility report

The classic form of written communication for the corporate treehugger is the annual corporate responsibility report. The production of these reports has spawned an entire industry for communications agencies and a substantial number of jobs for corporate treehuggers. Over the last ten years, the number of corporate responsibility reports filed with the Global Reporting Initiative has increased by 300% and most of the largest firms as tracked by the Fortune 500® now file some form of responsibility report.[68]

There is a huge amount of information and detail in these reports and every year the level of disclosure is increasing. Both Intel and AMD (my former and current employers) started publishing public reports in 1995, long before it was common practice. Back then, these were known as environmental reports, which went into the details of air, water, and waste emissions as well as any voluntary efforts to improve the environment.

Over the years, stakeholders who wanted to know more about the company's operations and policies demanded additional information. In the late nineties, the Boston-based nonprofit Ceres started the Global Reporting Initiative (GRI), which issued its first GRI Sustainability Reporting Guidelines in 1999. As of this writing, the GRI is on its third major revision of the guidelines (the GRI G3 and G3.1) and is well on the way to issuing a fourth. The GRI guidelines have become a de facto international standard for the issuance of "non-financial" or "environmental and social governance (ESG)" reports. More recent versions of the guidelines incorporate an "application level," which is a self-assessment of the degree to which you answered all the questions in the guideline. Letter grades (either A, B, or C) are assigned based on how many of the "core" and "non-core" indicators are covered in your report. You can get a "+" added to your grade if you have your report audited by a third party.

Each GRI guideline revision has added additional indicators that should be included with your report. There are now approximately 125 different data points in the GRI G3 guidelines covering detailed information in the following areas:

- Vision and strategy
- Organizational profile
- Report profile
- Governance
- Economic performance

- Environmental performance

- Social performance

- Human rights

- Society

- Product responsibility

Obviously, it is a significant amount of work to collect, verify, and report all of this information. In addition to reporting on the GRI information, most corporate responsibility reports are also a significant marketing document for the company. This means that you must develop storylines, create graphics, acquire images (with copyrights), and pull all of this together into a compelling report that complies with your company's brand guidelines. This is an important opportunity because CR reports are not only a wonderful way to represent your company's good work to he outside world, but also a source of pride for your company's employees across the globe.

Annual CR reports are an important element of any corporate responsibility program. Given the large scale and scope of these reports, it is very likely that you will be involved in the creation of these reports at some point in your career as a corporate treehugger. Below are a few tips for planning and creating a corporate responsibility report.

The hidden value of corporate responsibility reports

At the 2010 Ceres conference there was an award ceremony for the year's best CR reports. In accepting the award in the small and medium business category, the representative from Seventh Generation asked the audience how many people had read its report. Very few hands went up. Then he asked how many people had read their company's report. Almost all the hands went up. The next award winner was Ford Motor Company. During his acceptance speech, the Ford representative said that the process of pulling together its sustainability report was instrumental in driving programs within Ford. He said

→

the reporting deadlines impelled departments within Ford to launch programs and/or collect data that would likely not have happened without this forcing function.

These two award-winning companies were publicly admitting that the main value in producing their CR reports was not transparency (i.e., a window on company operations), but a driver for company awareness and action (i.e., a mirror for employees to evaluate and enhance their corporate responsibility actions). Few people will invest the time to read a CR report unless it has direct impact on them. In the age of instant information, if we want data on a particular issue, we don't sit down to read a report, we "Google" or "Bing" it so we can read a summary. This means that perhaps only a handful of people have read your CR report cover to cover, and most of those people are probably company employees.

Determine your format

The first step is to decide the format for your report. In the early days, most companies published printed reports and physically distributed them to their stakeholders. Since printing thousands of copies of a lengthy report can be a waste of resources, most companies have evolved to other formats. The most common format in use today is to develop a corporate responsibility website and append the full annual report as a downloadable file.

The website may capture abbreviated versions of the same content in the annual report, but can also include unique and timelier content such as blogs, tweets, or videos. The full annual report is like a complete record of your company's performance for the past year. Because the scope of these reports has grown so large (along with the page count), some companies have developed a "report builder" function that allows the stakeholders to select only the sections of interest for download.

In addition to the Web and downloadable version of the annual report, many companies also develop a short summary in a printed brochure to use as a handout with certain audiences. Once you determine which combination of these formats you will use, you can outline a project plan. Each of the formats involves a different set of specialties (Web, graphics, layout, video, print, copywriting, etc.) that will have to be brought into the project team.

Start early

These reports are a massive effort. For example, AMD publishes its report in early May. The team assembles to start the process in November. As the project unfolds, there are upwards of a dozen staff working on the report and a dedicated, full-time team leader. While the company has not counted the man-hours dedicated to this report, all told, I would estimate it is well over 1,000. If your company has done a report in the past, you can look back at the business processes from the prior reports and simply update these. Start by asking your internal stakeholders about what worked well and what did not. For example, you may find out that your last report was issued late, had data gaps, or ran into approval or branding problems. Take stock of the issues and begin planning your schedule, budget, and process improvements early on. Most reports are issued in the first quarter or early second quarter of the company's fiscal year in order to coincide with the company's annual meeting. To meet this schedule, you have to start planning early in the fourth quarter. If your company is on a calendar year, this can be difficult because the fourth quarter is full of other demands as well as holidays. Even if you can only get the team together once or twice, it is far better to have started in the fourth quarter than showing up in January and trying to sort it all out.

Form a team

Start planning your CR report by outlining roles and responsibilities for each of the major internal stakeholders. In Chapter 2 we talked about "leading through influence" and "reading the system." These skills become paramount here, because you will be asking people to do a lot of work for this report, and they need to see the value that justifies their effort.

Collecting the data you need is the first step. If your company is already in the rhythm of doing an annual report, it is likely that you will already have a network of data providers identified from all of the relevant departments. In these cases, the effort can be described as an "update" of last year's report.

As you contact colleagues to request the needed data, you will likely encounter reactions ranging from eager to reluctant. Ideally, the departments providing data for the annual report will own their section and take pride in telling their part of the CR story. If this is the case, your role is to manage the overall structure and production of the report. Most likely, it will be a mixed bag: you will end up developing almost all of the content for some sections and very little for other sections.

When working on a corporate responsibility report, you will likely encounter "silo behavior," where people only relate to the issues within their own department (silo) and don't readily share information with others. When you encounter reluctance (e.g., ignored calls, unanswered e-mails, missed deadlines, or blown-off meetings), set a meeting with the leaders of the department along with your own management. Discuss why the report has value to all involved and strive to define a point of contact and a deadline for delivering the needed information. If this does not work, you can continue to escalate up the management chain or simply go forward without the information. It is probably better to take a hit on your GRI scores than cause a feud between your group and another business team.

Start with the GRI

Many of the reports I have worked on did not follow this simple rule: *start with the GRI*. What happened in these cases is that we scrambled after the report was essentially done to figure out how we could fill all of the gaps in the required GRI data. It is far better to start with the GRI guidelines as the format for the request to your data providers.

Get the latest version of the guidelines and split them up into sections that are relevant to your company's structure. If you have reported in the past, create a table that shows the data from the last report and adds a column for them to fill in for the current reporting year. In the same request, you should ask your data providers for stories and anecdotes about major changes or achievements in their department. For example, the community affairs department will report the number of volunteer hours and can also provide you with great stories of individuals or groups who have volunteered in their communities. In some cases, these departments will want to draft these stories themselves; in other cases, you can interview people to gather the stories.

At the end of the report, you should publish a GRI index. This is an index of each GRI indicator, whether you reported it fully, partially, or not at all, and where it can be found in the report. From the index, you can derive your grade or "application level" and publish this as well. Finally, if you use a third-party firm to assure your data, you will include a statement from that firm detailing its assurance process and findings.

Create a master plan

As you can tell from the pages above, the CR report is a major endeavor. It cannot be done without assembling a team of people to acquire the data, create the creative layout, develop graphics, Web programming, brand compliance, images, copywriting, editing, and legal review, etc. As you set out your project plan, you will need to figure out where each of these services will come from. It is like being the general contractor for building a house – you will need to arrange for the carpenters, plumbers, and roofers, etc. to perform their trade in the right sequence so that the whole project comes out on time and on budget.

Because the corporate responsibility report is a ubiquitous communication tool and can be the top priority of the corporate responsibility department, condensed below are ten steps to help plan your corporate responsibility report:

1. Decide on deliverables (Web, PDF, brochures, printed reports, etc.) and the deadlines for each

2. Outline the sections of the report based on the GRI guidelines (specifically create a table that accounts for where each of the GRI indicators will appear in the report)

3. Define your network of data providers

4. Define your network of service providers for copy, Web, layout, images, graphics, and print as appropriate

5. Create a master schedule with milestones for:
    - Data and story acquisition
    - Copywriting and editing
    - Web design and programming
    - Layout, images, and graphics for Web, PDF, and print
    - Reviews by appropriate executives
    - Reviews by legal (and brand conformance, if needed)

6. Define your budget and allocate resources to each function

7. Define a schedule for each function that sequences the activities by their dependencies (e.g., the raw data is needed before the story and copy can be created; or management review must occur before legal review, etc.)

8. Schedule a kickoff meeting with the full team to get buy in to your schedule and budget and make any needed changes

9. Schedule team meetings at a regular cadence throughout the project

10. Make sure to recognize the team when then project is completed

In many cases, companies outsource some or all of these functions to a communications agency. Even in these cases, you will need to develop a plan to gather all of the information needed for the report, provide project oversight, and manage all of the needed internal reviews.

Beyond the corporate responsibility report

It's May. You just spent a third of your year working on the annual CR report. Your reward? More communications . . .

In the Internet age, annual corporate responsibility reports are too slow to be sufficient as your only means of communication. While the annual report is a critical document to establish the record of your performance and goals, your stakeholders want more frequent and timely communication.

Not too long ago, corporate communications mainly consisted of professionally written press releases that were reviewed by subject matter experts, executives, and legal counsel then issued to wire services and relevant publications. These press releases could take days or even weeks to produce. Today, most corporate communications departments have one or more Facebook fan pages, Twitter accounts and a few blog feeds that are updated in real time. Continuous and instant corporate communications are now a fact of life.

The expectation for continuous fresh content has rewritten the rules for corporate communications, including corporate responsibility. Most leading CR departments are now producing an annual report *as well as* a variety of additional communications that are issued on a more frequent cadence:

Newsletters

Several leading corporate responsibility departments are issuing newsletters that are published more frequently than the annual report. For example, AMD issues a "corporate responsibility update." This publication is an e-newsletter format issued in the summer and fall of each year (the annual report is typically in the spring). AMD uses this publication to update its stakeholders on progress across a broad range of issues, and it also features an in-depth look at a specific CR program. These updates are a great way to rotate through the initiatives within your program's scope, not only to keep your stakeholders informed, but also to engage your business partners by spotlighting their activities and accomplishments.

Another advantage of issuing more frequent updates is creating the distribution list of interested stakeholders. Along with your Twitter and Facebook followers, this list forms part of the stakeholder community for your program. Avoid spamming this community by allowing them to both opt in and opt out of the list.

Blogs

Many leading CR sites feature a blog feed. Blogs are both a blessing and a curse. They are a blessing because you can issue information on a frequent basis, cover a broad scope of issues, and feature bloggers from the major functions within your company. Blogs can become a curse because of the amount of work it takes to keep them fresh and on message.

Here are a few steps to manage a corporate responsibility blog:

- **Identify your bloggers.** Make the rounds of the major contributing departments (EHS, HR, legal, etc.) and ask them to identify people who will write blogs to become official bloggers for the CR blog feed. Set up a "meet the bloggers" link with a headshot and quick bio and provide each blogger with either editing or access to ghostwriting to ensure that you get consistent quality and delivery for your blog feed

- **Calendar your blogs.** Develop a calendar of all the topics and bloggers that you can forecast and assign them an owner and a due date to make sure that the content is delivered

- **Establish a process.** Set out a review process for your blog posts including your communications and legal team and establish a regular meeting to review the blog calendar (suggest bi-weekly) to stay on track and adjust to new developments as needed. Although this will not be appropriate for every topic, try to post your blogs on other sites in addition to your corporate website. For example, GreenBiz.com, Sustainable Life Media (sustainablebrands.com/sustainablelifemedia), 3blmedia.com, csrwire.com, and others are great outlets for good content. Posting to a well-known CR site will guarantee that you will get more visibility. Be selective on the blogs that you promote for external sites – they should lean toward more cutting-edge themes as opposed to promoting your company's programs.

Social media

Twitter and Facebook have already changed corporate communications in profound ways and these changes are still rapidly evolving. On the plus side, social media provides a forum for an ongoing dialogue with a self-identified group of people who are interested in the company's messages. On the flipside, social media can be a black hole for fresh content with little time for preparation or review. The analogy is the ravenous carnivorous plant Audrey II in the movie *Little Shop of Horrors* that keeps screaming, "Feed me, Seymour!"

- **Coordinate with corporate communications.** Work with your corporate communications team to set up your Twitter handle and your Facebook site. It is likely that your company already has a social media presence. It may or may not make sense to house your corporate responsibility messages on the company's main Facebook site. Unless you work for one of the "epiphany" companies described earlier (companies that are founded on sustainability as a central tenet), your audience is likely to be somewhat different than the audience for your mainstream corporate messages. This is not to say that you should not cross-post these messages, but in many cases, it makes sense to have a separate social media presence for the corporate responsibility function. Your followers will appreciate a more focused stream of communications about topics that are more relevant to them

- **Develop a process.** You will need to assign certain people to have access to your social media accounts and develop a system for accountability. While you will not always have time to review them, every tweet and Facebook post is a voice representing your company. The best way to manage in this paradigm is to restrict access to only a few accountable people who have been trained in media relations, understand corporate responsibility

content, and are in the loop for corporate position statements
(in some companies managing social media channels has been
defined as a full-time role)

- **Do it daily.** Static social media defeats the purpose and will work
  against the credibility of your program as your followers begin
  to check out. If you are going to enter the social media world,
  you have to keep up a steady stream of communications. To do
  this, the people you have assigned to update your social media
  accounts should be the "plugged-in types" (typically the ones
  who send a lot of e-mails with interesting links and "FYI" in the
  subject line). Once recruited, these people should be running
  TweetDeck or HootSuite (these are programs that aggregate your
  social media accounts such as Twitter, Facebook, LinkedIn, etc.)
  and posting items of interest at least daily, if not several times
  a day. These folks should also monitor the feeds on hashtags
  related to your company, your priority topics, and any trending areas of interest. To do this does not require a lot of original
  content; retweets of interesting posts and articles are a great way
  to feed the pipeline by using nothing more than "check this out"
  as a lead

- **Monitor your comments.** The world of social media is a two-way
  conversation. Your followers will let you know what they think
  on a real-time basis. Continually monitor comments on your
  blogs, Facebook posts, direct mentions, and retweets on Twitter.
  It is a judgment call on when to respond to comments but avoid
  being drawn in by the "haters" who live on the Web and rou-
  tinely post negative comments. The trick is to be able to discern
  real input and trends from the crackpots and spammers

- **Special events.** I recently watched a "CSR chat" by Best Buy
  (U.S. electronics retailer). The event was live-streamed video on
  the Web with a very active twitter hashtag. The Best Buy folks
  were able to tell their story and respond to questions from the
  Twitter feed. The comments and dialogue from the Twitter followers were universally positive with one tweet exclaiming that

the event raised the bar for sustainability communications. Any time that you can involve your followers in an interactive forum like this, you get a rich mix of feedback and more people learning about your CR story. Other companies have held contests or created interactive applications as a way to connect with their consumers on corporate responsibility themes. For example, Timberland's Earthkeepers Virtual Forest promotion on Facebook allows fans to plant a virtual forest and watch it grow. The promotion was so successful that Timberland had to shut it down temporarily when it ran short of tree seedlings

- **Stay current.** Just a few years ago, Facebook was only for college kids and Twitter did not exist. Now, every communications department has a social media strategy dealing with these channels

- **Add a CR-focused approach.** Twitter, Facebook and Google+ are terrific social media channels for all interests, but there are a growing number of online communications channels that are exclusively focused on corporate responsibility. Triple Pundit, Triple Bottom Line (or 3BL, which recently acquired JustMeans. com), CSRwire, GreenBiz.com, Environmental Leader, Vault. com, Treehugger.com, and ethicalcorp.com are all examples of the special interest channels that you can use to ensure your messages are shared within the CR community

Audio and video

Audio and video files are becoming increasingly important communication vehicles. Short video clips and podcasts can be shared through your social media accounts and websites. Many companies also have a YouTube channel to feature their latest video clips.

Obviously, audio and video are more technically challenging and require specialists who can produce, tape, and edit content into a compelling piece. A few tips for working with audio and video are:

- **Manage the content.** With some experience, you can become adept at the video medium so that you can manage the message. Don't feel intimidated if you have not worked in this medium before. Like other formats, it is essential to convey your message to the target audience in a clear, concise, and memorable way. With video and audio content, you will need to assemble an outline or script with short soundbites from subject matter experts, B-roll (background video with voiceover), and professional voiceover narration. If your company has a studio, you can work with them or hire an external communications agency to produce your video

- **Keep it short.** Long videos or podcasts of corporate communications can be boring. Less is more when it comes to this form of communication. You need to define one or two key messages and stay focused on these. The rule of thumb is to shoot for about two minutes – much longer than two minutes and you risk losing your audience

- **Keep it interesting.** Avoid making videos that are overly promotional about your company – these can come across like selling timeshares. If you want to capture hearts and minds, stay humble and focus on topics that have general interest. You will be able to weave in your company's message but it should be within the context of an issue that has broader appeal. For example, if your company is working on a controversial issue like eliminating sweatshop labor in your supply chain, frame your company's actions around general information on the topic

Pulling it all together

Throughout this book, I have pointed to communications skill as one of the most important attributes for the corporate treehugger. Today's corporate responsibility group is like a mini-newsroom. You must be up to speed on the latest developments internal and external to your company and constantly communicating in multiple formats and venues.

You have to not only be able to create the story, but also be competent enough to tell the story in a compelling way.

A common communication mistake is to over-communicate. "If I had more time, I would have written a shorter letter": this quote is a fantastic reminder that it takes preparation and planning to deliver concise, compelling messages that will resonate with your intended audience. The take-home message in all of these tips is that communication in all of its forms is a defined, learnable, and essential skill for the corporate treehugger. Like mastering any skill, it takes practice, coaching, perseverance, and patience. Mastering this skill, however, pays huge dividends. Communications is a portable skill that can be put to good use in almost any career choice and is arguably the single most important attribute for your success.

10

Stakeholders and investors

This chapter outlines how to identify, prioritize, and engage with external stakeholders to add value to your corporate responsibility program.

Feedback is a gift (source unknown).

When I started my career in environmental management in the mid-1980s, the term "stakeholder engagement" did not exist. Today, there are businesses built around providing your company with this essential aspect of your corporate responsibility program.

There are several schools of thought about engaging stakeholders in your program, and like so many things in the sustainability/corporate responsibility field, the definitions can be vague and imprecise. Let's start with a definition of stakeholder engagement:

> Stakeholder engagement is a formal process of relationship management through which companies or industries engage with a set of their stakeholders in an effort to align their mutual interests, to reduce risk and advance the triple bottom line — the company's financial, social, and environmental performance.[69]

At its essence, stakeholder engagement is really about getting *input on your program goals, progress, and plans.* Because the reputation of your company is a hallmark of success for your corporate responsibility program, it makes sense to identify the stakeholders who influence your company's CR reputation and invite them to participate in your program. In the process you will build relationships with these individuals and groups and gain valuable insights for shaping your program. This chapter will delve into these rules of engagement and give some specific examples of effective practices.

There are several ways to engage with your company's stakeholders, ranging from somewhat passive means (e.g., suggestions received on your company's website) to a formal standing advisory panel. While there are a myriad of variations for engaging stakeholders, this chapter will cover practical advice for *formal engagement* (e.g., stakeholder panels) and *informal engagement* (e.g., consulting with a few external contacts) that you can adapt to your circumstances. While your company's employees are an important stakeholder group, this chapter is solely focused on stakeholders who are external to your company (employee engagement is covered in Chapter 11). Also, I deliberately distinguish the socially responsible investment (SRI) community from other stakeholders because there are some unique aspects of engaging with this segment that warrant additional advice covered later in this chapter.

Who are stakeholders?

A stakeholder is any group or individual that has the ability to impact and/or that may be impacted by your company's operations and/or policies. Stated another way, a stakeholder is an entity that has a legitimate social, economic, political, or environmental "stake" in your company's activities. The term "stakeholder" was once more narrowly defined as shareholders, regulators, employees, and customers. Today, as a result of changing expectations, stakeholders comprise a much broader set of constituencies that must be defined on a case-by-case basis for the particular circumstances of your company.

Stakeholders may include:

- **Ownership interests.** Investors, partners, shareholders, analysts, and ratings agencies

- **Customers.** Business customers, consumers, and consumer advocates

- **Employees.** Current employees, potential employees, retirees, and labor unions

- **Value chain and operational support.** Suppliers, contractors, and service companies

- **Industry.** Industry associations, industry opinion leaders, and competitors

- **Community.** Residents near company facilities, chambers of commerce, resident associations, schools, community organizations, spiritual communities, special-interest groups, and indigenous peoples

- **Civil society.** NGOs, activist groups, charitable associations, and clubs

- **Government.** National or federal policymakers, state policymakers, local policymakers, regulatory and tax authorities, and customs officials

- **Multilateral organizations.** Examples include the United Nations, World Bank, and the International Finance Corporation

Formal stakeholder engagement

Formal stakeholder engagement involves establishing a committee that you will work with over the long term, either on a particular project or as a standing advisory board to provide general input for your program. This type of engagement carries the most overhead in terms of governance procedures, ground rules, and operations, but it can also provide the biggest benefits to your program. The principal benefits of operating a formal stakeholder panel stem from the relationships you will build by working with a committed group of people over a longer period of time. The stakeholders will gain a deeper understanding of your company and the issues you face. This understanding, coupled with their diverse perspectives, will provide insights and guidance that can help guide your program. Also, if your company encounters difficult issues, the stakeholders on your panel will be well positioned to comment on your behalf because they will be invested in your program and knowledgeable about your strategies.

Project XL

In the mid-1990s, I was involved in one of the early environmental programs based on formal stakeholder engagement. The program was called Project XL, which stood for "eXcellence and Leadership," and was a cornerstone of then Vice President Al Gore's "reinventing government" initiative. The basic idea was to develop a process to improve environmental performance while reducing government bureaucracy. The projects that were authorized under this program featured formal stakeholder committees to engage in the decision-making process and oversee the performance. Because this project broke new ground in the arena of stakeholder engagement, it is a useful model to extract a few salient tips for formal stakeholder engagement today.

Intel submitted a proposal focused on a facility still under construction at the time, called Fab 12 in Chandler, Arizona.[70] The essence of the proposal was that all of the facility's air emissions would be totaled up and reduced so that the total pollution from the facility would be significantly less than the regulations would require. In exchange for this added level of environmental protection, Intel would not have to seek additional permits each time the manufacturing process changed. The project included a stakeholder panel that would be intimately involved in defining the final agreement as well as overseeing implementation.

To identify the stakeholders for this panel, Intel started with the existing community advisory panel (CAP) that was made up of neighbors, local activists, and local regulators. The CAP appointed members from their ranks that had an interest in environmental issues to the stakeholder panel. In addition, Intel made sure that each of the federal, state, and local regulatory agencies with a relevant oversight role was represented on the panel. Finally, because this was a precedent-setting project, Intel invited some national NGOs to participate as well.

As the group came together, they had to define their roles and responsibilities. There were long debates over whether the stakeholders on the panel would need to be in full consensus (unanimous agreement) or whether a majority vote would suffice. Consensus was considered problematic because it essentially gave veto rights to each panelist. I recall my boss at the time, Larry Borgman, had a phrase that encapsulated stakeholder roles, which he dubbed "the four Vs." This stood for "View, Voice, Vote, Veto," and his belief was that stakeholders should have the first two but not the second two Vs.

The panel eventually decided to use a "group consensus" process to reach decisions. This meant that each of stakeholder *categories* had to agree. In other words, all of the panelists were assigned to groups, such as NGOs, local residents, regulators, and Intel, and each of these groups got one vote. If any group objected, the vote would fail. This allowed consensus to move ahead even if one member of the panel objected but their group was in favor.

The ultimate success of this project was due, in no small measure, to our very talented facilitator, the late Chuck McLean who, despite constant veto threats and charged emotions on this project, brilliantly led the team to consensus on every tough issue.

The stakeholder panel broke into sub-teams that met at least monthly to work through each aspect of the permit. Because the issues we were discussing in these groups were technical, Intel funded independent technical analysts that were chosen by the stakeholders to help them understand the concepts and thus be more fully engaged. Intel also contributed funding to cover the expenses for some of the stakeholders.

This stakeholder panel was ultimately successful in producing a final agreement that allowed Intel to operate without changing its operating permit, significantly reducing the facility's environmental impact. Because the permit capped the total emissions from the site, Intel was motivated to develop innovative ways to keep emissions under that threshold so that they could continue to expand their operations without triggering time-consuming environmental permitting processes.

The panel stuck together for more than a decade after the project was approved to oversee implementation. During that decade, Intel was able to build two additional factories (at a capital expense in excess of $6 billion) on that site without ever changing the permit negotiated under Project XL. This level of flexibility is unprecedented for environmental permitting and resulted in significant savings in time and money for Intel.

Establishing a stakeholder panel

Below are a few important principles for establishing and operating a formal stakeholder engagement panel:

Define the reasons to engage

Not every company, program, or project needs a formal stakeholder advisory panel. Operating these panels is time and resource intensive and thus you should use them only in certain situations. Circumstances are appropriate for formal engagement with a stakeholder panel when:

- Your company is developing a new strategy and/or goals

- Your company is changing its business model or building a new facility

- You want to significantly revise your corporate responsibility programs

The common element in these cases is that your company is experiencing significant change, seeks input, and is open to act on the advice. If you are more interested in getting the credibility that comes with engaging stakeholders or just testing the waters to get reactions to your programs, a formal engagement model is not recommended and may backfire.

Find a facilitator

I highly recommend starting with a talented facilitator. When seeking independent perspective and opinion, a third-party facilitator can help guide the process so that you get true diversity of perspective and influential opinions. I have used Business for Social Responsibility (BSR) while at Apple and, more recently, Ceres at AMD – both are excellent at establishing and facilitating stakeholder engagement processes.

Identifying stakeholders

Often called "stakeholder mapping" the 2x2 matrix in Figure 6 sets out the generic process you can apply to identify your stakeholders. Again, this is a process that is best conducted with a third-party facilitator because it takes you out of the role of "hand-picking" your panel. The people and groups in the top-right corner of the matrix – those who are both *interested and influential* – are the best candidates for a formal panel. This is not to say that other segments of the matrix are unimportant. The matrix is only a guide to establish priorities for how to interact with different groups of stakeholders.

Stakeholder mapping works well when you have already identified the "material issues" in your corporate responsibility strategy (see Chapter 3) because your facilitator will be able to match the stakeholder interests to these issues. Of course, you may also engage stakeholders to guide you in the process of selecting material issues, but the makeup of the group may change based on the issues selected. Typically, your team will work with the facilitator to put names of people and groups into the matrix in Figure 6.

Figure 6 Stakeholder mapping matrix

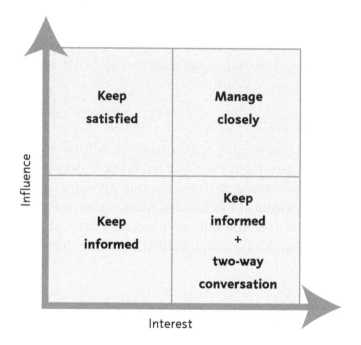

The matrix illustrates four categories of how your company might choose to engage with various stakeholders. There is a continuum of engagement methods that ranges from "ignore" to "partner," but for this example, we are looking for the people/groups to join an advisory panel (the "manage closely" quadrant of the matrix). In many cases, the stakeholder groups to invite to the panel will be obvious. For example, if your company is frequently targeted by environmental activist groups for its use of packaging materials, you would want to select one of the more influential, yet reasonable, groups focused on packaging issues. The matrix provides a tool for your team to rank the various groups by their level of interest in topics that concern your company and their level of influence over those topics.

While some stakeholder groups are ideal candidates for your advisory panel, you should not lose focus on other organizations. For example, Greenpeace is highly influential on packaging issues but is more interested in a competitor than your company. In this case, it makes sense to put Greenpeace in the "keep satisfied" quadrant to avoid becoming its next target. Engagement with the stakeholders in this category could include holding a regular dialogue with them to understand their issues and report on your progress on these issues. Since Greenpeace is focused on preserving the rainforest, your dialogue should focus on your company's policies and programs to eliminate packaging made from rainforest materials.

For the stakeholders that are not as influential as Greenpeace, but have demonstrated genuine interest in your company's CR programs, it is beneficial to keep them informed and listen to their input. For example, the neighborhood association near one of your company's facilities has repeatedly contacted company management about noise issues. While this group's narrow interests and low influence would not make them a great candidate for a standing advisory panel, it is still very important to have an active dialogue with stakeholders that are interested in your company. Ignoring these groups risks alienating them, which could lead to protests, or worse. By engaging with interested groups you can not only avoid alienation, but also create ambassadors for your company.

In addition to this initial sorting of stakeholder interest and influence on key topics, your team should assess the posture of stakeholders toward company engagement. While some stakeholder groups are interested and influential, they might be poor candidates for an advisory panel. For example, groups that are established to campaign against companies, such as Greenpeace, are not likely to affiliate themselves with a company advisory panel, nor would their participation be all that useful in this context. These groups are far more comfortable working from a distance so that they can maintain their independence and highlight alleged company misdeeds in the press.

Other factors to consider when selecting stakeholders include:

- **Issue relevance.** Does the stakeholder focus on the issues relevant to your company?

- **Willingness to engage.** Does the stakeholder have the experience, time, and inclination to engage with your company?

- **Influence.** Does this stakeholder reach a broader audience? This factor is often the critical decision point

- **Innovation.** What is the stakeholder's track record for creating new approaches or contributing to partnerships?

- **Location.** Is the stakeholder focused in the geographic areas of importance to your company?

Define the issues for engagement

Start the process by defining the key issues for review and discussion. These issues should be important for your company and for the stakeholders on the panel. Make sure that these are issues for which you truly want and can use feedback. For example, at the time of this writing, AMD recently established a new set of material issues and is working on the development of strategies for them. Working with Ceres, AMD is establishing a stakeholder panel to review this work and advise us on these strategies as well as ways that AMD might measure and improve its overall corporate responsibility performance.

Establish governance and ground rules

In the Project XL example above, Intel used a very structured governance process that was appropriate for that project. Typically, the rules of engagement will not have to be as defined or rigid as this example. In most cases, the sponsoring company is seeking advice and it is clear to all involved that the ultimate decision for acting on the advice remains with it. Even so, it is useful to define the purpose, scope, and objectives of your panel in a **charter document**. This document should outline the following elements:

- **Purpose and scope.** This section of the charter document outlines the scope of the engagement with the panel. For example:

the panel will focus on advising the company about corporate responsibility strategies. It should also state that the panel serves as an advisory body and the company is under no obligation to act on advice from the panel (again, it is best if you demonstrably take actions based on the panel advice, but this provision in your charter can help align expectations). You should also include language in this section of the charter that makes it clear that no confidential information will be shared with the panel to avoid any implication that this is a group of "insiders" (this is especially important if your panel includes investors)

- **Governance.** Typically, the panel will operate by consensus. When the panel is established as an advisory group, a governance provision is somewhat unnecessary since all opinions are welcome. If the panel will have some decision-making authority, it is essential to establish the ground rules for reaching a decision (e.g., consensus, simple majority, group majority, group consensus, etc.) in this section

- **Confidentiality.** This section sets out how the identities and opinions of the stakeholders will be shared beyond the panel's operations. In most cases, panels operate by the Chatham House Rule, which states: "Participants are free to use the information received, but neither the identity nor the affiliation of the speaker(s), nor that of any other participant, may be revealed." There may be cases where you and/or your stakeholders want to reveal information from the panel – for example, if you chose to publish panel opinions in your corporate responsibility report. In these cases, obtain all panelists' consent to disclose in writing before proceeding

- **Funding.** It is important to define the extent of financial support that your company will provide to stakeholders for their participation. Financial support is negotiable, but is typically provided for the nonprofit participants for items such as travel costs and independent technical assistance

- **Facilitation.** Make it clear that a third-party facilitator will be utilized to manage the panel operations at the cost of the sponsoring company

- **Terms of membership.** This section sets out the length of engagement and the expectations for the number and duration of meetings as well as any deliverables. This section should also outline a code of conduct that members are expected to follow and the process for releasing a member from the panel. For example, a member could be released if they violated the confidentiality terms

Informal stakeholder engagement

There is a range of ways to get input from stakeholders that are less formal than creating a standing stakeholder panel. Although these methods are less formal, it still makes sense to go through a stakeholder mapping exercise to identify the people and groups that you want to contact. The main difference between formal and informal engagement is the depth of the commitment. In the formal stakeholder panel the participants have made a long-term commitment to the company and issues. Informal consultations have no such commitment. The advantage of informal stakeholder engagement is that you can get feedback quickly and easily from a broad sample of people and/ or groups. The disadvantage is that the feedback may not be as well informed and you won't necessarily have the time to build the relationships and trust that come from a formal panel.

Below are some proven methods to engage with stakeholders short of setting up an informal panel:

Exercise your network

Look over your network of contacts and conduct a quick stakeholder mapping exercise. Select a few influential people and set up phone calls or one-on-one meetings to cover the issues in your program and obtain feedback. A good practice is to draft an interview guide and background document ahead of time and share it with the people you will call. Also, you should think through what you might offer them for their time. For example, if the results from interviews might be of interest to them, you should offer to share a summary.

Online forums

In the Internet age, it is almost impossible to avoid a continuous dialogue with a wide variety of stakeholders though social media. Just by monitoring and responding to the comments on your blog feed, Twitter, and Facebook accounts, you will get a sense of stakeholder opinions. You may also have a "contact us" feature on your CR website and report that will garner some stakeholder feedback (in my experience, however, this format generates more solicitations than real feedback). You can also develop more focused online forums to solicit feedback by setting up challenges, prizes, or contests on your website, Facebook page, or other external site. One of the more innovative examples of online stakeholder feedback is from SAP. The company published its CR materiality matrix on its website in a format that allowed users to move the issues between the quadrants on the matrix.[71] Its system records the selections by each person who interacts with the site.

Third-party research

This method engages a third party as a go-between for the company and the stakeholders. This allows the stakeholders to be more forthcoming with opinions. Depending on the situation, you can set up these interviews to allow the stakeholders and/or the company identity to remain anonymous.

Focus groups

These are similar to the formal stakeholder panel model discussed above with one important difference: the stakeholders have no continuing obligation after the meeting. Focus groups are typically issue-specific meetings where your facilitator is collecting opinions on an issue from a diverse group. It is wise to go through all of the steps you would in setting up a standing stakeholder panel – defining the scope of engagement, selecting participants, establishing ground rules, and selecting a facilitator – but within a more narrowly defined scope and schedule.

Engagement with socially responsible investors (SRI)

Socially responsible investors are a distinct category of stakeholders for your company's CR programs. The SRI community is a diverse mix of investment fund managers and analysts who are, in general, very sophisticated in their level of understanding of both your company's business model and your CR strategy. To understand how to work effectively within the SRI community, it is helpful to divide it into two categories: SRI analysts and SRI funds.

SRI analysts

These are the firms that collect and analyze your company's environmental, social, and governance (ESG) data and provide the results to investors. Sustainable Asset Management (SAM), Morgan Stanley Capital International (MSCI), Trucost, Sustainalytics, and IW Financial are some of better-known entities providing this service. While there has been consolidation in this field, there are still many SRI analyst firms that gather data for the investment community.[72]

ESG analyst firms also sell their research to provide the data for company rankings such as *Newsweek* magazine's "Green Rankings" and *Corporate Responsibility Officer Magazine*'s "100 Best Corporate Citizens" list. Each of these rankings utilizes a different and sometimes proprietary method for ranking company performance. The "Rate the Raters" study by the consulting firm SustainAbility does a nice job of exploring the world of corporate responsibility rankings and ratings.[73]

The important thing to understand is that each of these analyst firms uses a different and often proprietary screening method to evaluate your company. For example, one analyst may weigh water conservation more heavily than carbon management while another may put labor issues in the forefront. The bottom line is that there is no harmonized way to measure corporate sustainability, so each firm does it in its own way.

SRI funds

The number of investment funds that apply a social/environmental screen to company stocks is growing rapidly. The 2010 Social Investment Forum report states that there are now 493 SRI funds, which is more than double the 201 SRI funds that existed in 2005. Some of these funds have been around for quite a long time, such as Calvert Investments (founded in 1976) and Trillium Asset Management (founded in 1982), but new SRI funds are popping up all the time. Also, many of the large public employee retirement funds such as TIAA-CREF and CalPERS use some form of ESG screening for a portion of the funds they manage.

From my experience, there is little consistency in how SRI funds screen companies on ESG issues. They may buy research from one or more of the analyst firms or use their own screening methods or some hybrid. Many of these funds also have blanket prohibitions against buying stock in certain firms depending on their business. For example, many of the religiously based investment funds will avoid companies involved in tobacco, alcohol, or weapons.

It is likely that members of the SRI community will be participants in the various forms of stakeholder engagement discussed above, but because the SRI community can influence your company's stock, there are a few additional methods to interact with them discussed here.

The SRI world is complex. There are a myriad of socially screened mutual funds, ESG investment analysts, and responsibility ratings systems. For example, the 493 U.S. investment funds in the Social Investment Forum report accounted for $569 billion in total net assets.[74]

Each SRI fund utilizes its own screening system to decide which companies it will include in its portfolio. As mentioned above, some funds simply screen out companies involved in alcohol, tobacco, or weapons. Other funds have their own in-house analysts that screen investments based on their investors' social or environmental priorities, or use the research from ESG analysts described above to select stocks.

If you get the impression that working with the SRI world is complex and time-consuming, you are correct. However, there are huge benefits that you can achieve for your company by engaging with this stakeholder group. The primary benefits are:

Increased investment

The Social Investment Forum estimates that, as of 2010, the total professionally managed assets following SRI strategies amounts to $3.07 trillion and is growing at a faster pace than conventional investment assets. This report estimates that SRI assets grew at 380% from 1995 to 2010, while conventionally managed assets grew 260% during this same period. In addition, SRI investments grew during the 2008–2009 financial crisis while conventionally managed assets remained flat.[75] Based on these trends, some corporate investor relations departments have identified SRI funds and analysts as important stakeholders.

Lower volatility

SRI-managed funds are dominated by large institutional investors such as state or city employee retirement funds. These groups tend to be value investors rather than market-timers that move in and out of stocks quickly. These are the types of investor that appeal most to your investor relations department because they hold stock for a longer period of time and can lower the volatility of your company's shares.

Managing shareholder advocacy

For many years now the SRI community has collaborated on developing and filing shareholder resolutions on ESG issues. If your company
is publicly traded in the U.S., any shareholder with $2,000 worth of
stock can file a resolution. The resolutions can be on almost any matter involving the company and, with few exceptions, must be included
on the overall "proxy ballot" (where all shareholders get to vote on the
resolution – one vote for every share they own). Decades ago, social
activists discovered the shareholder resolution process as a way to
raise awareness within companies on a wide variety of issues. Often,
the SRI community will develop several standard resolutions that are
filed by shareholders of targeted companies. For example, in 2011, a
series of resolutions were filed that asked companies to assess their
human rights policies in countries outside the U.S. where their business operations are hosted.

The Social Investment Forum Foundation reported that more than
200 institutions filed ESG shareholder proposals between 2009 and
2010.[76] Typically these proposals ask the shareholder to vote on an
ESG issue that the filers would like to see the company address. For
example, Apple received a filing that sought to require the company
to issue a public sustainability report (as opposed to the report Apple
issued solely on supplier responsibility).

Each year, the number of these shareholder resolutions increases
and the number of votes they attract increases as well. By working col-
laboratively with the SRI community you will be in a better position
to understand which issues are likely to be included in shareholder
resolutions and possibly avoid a filing by proactively explaining your
company's performance on these issues.

ESG proxies, or shareholder resolutions, may demand that your
company disclose its carbon emissions or endorse human rights principles. If you encounter an ESG proxy, you can often negotiate with
the proxy filers (usually an institutional investor in your company)
and, by explaining your programs or agreeing to take additional steps,
you may be able to get the proxy withdrawn before it is added to the
shareholder ballot. After negotiating on several ESG proxy issues at

two companies, my experience is that working with the proxy filers is a far better process than putting the issue to a shareholder vote. While the company will usually win the vote, the ballot will have to include the issue and your company's voting recommendations – all of which have to run through a lengthy review process and can be controversial. By engaging in an open dialogue with the filers, you can build trust and mutual understanding. It is likely that you are working on the same issues in the filing and, with some internal negotiations, may be able to rearrange priorities and satisfy their concerns.

The SRI 'road show'

Investor road shows – where your company's executives travel to meet with a number of influential investors and analysts – have been a common practice in the investment world for some time. This format has also become an annual or semi-annual event for many corporate responsibility leaders. The SRI road show is simply a series of meetings with SRI fund managers and analysts, the purpose of which is to tell your CR story and hear their feedback. Your investor relations (IR) department, in conjunction with your CR team, usually sets up these meetings and agrees on the speakers and the key messages. Over the course of a few days, a team from IR and CR will meet with several fund managers and analysts in their offices – in the U.S. this means going to Boston, New York, and Washington, DC. Your company might also cover the SRI firms in the European market. The Asian SRI community is still in an early stage, but growing rapidly.

On my last SRI road show, we conducted nine meetings in two days with most of them scheduled back-to-back. The typical presentation starts with an overview of company financials, strategy, and outlook from the IR department representative. The back half of the presentation is focused on the corporate responsibility story. In my experience, the SRI community is completely focused on the CR section of the presentation. If you are ever on one of these road shows, expect that you will be on center stage.

The structure of the CR portion of the presentation should cover the background for your program, strategy, goals, performance indicators, and any hot topics. For example, during AMD's last SRI road show we spent a substantial amount of time discussing a relatively new and sensitive issue: conflict minerals. This issue refers to the use of minerals originating from conflict areas in Central Africa. The profits from mining and trading in these minerals have funded armed groups responsible for mass killings and rapes in the region.[77]

Whether your role is to present the CR story or if you are helping to develop the content for these meetings, it is best to set a schedule in advance to go over the presentation and hold a dry-run session with the team. In your dry-run session, assign someone to play the role of the investors and analysts and ask hard questions about the issues in your presentation and other issues that you may not anticipate. Remember, this is a sophisticated audience with years of experience reviewing corporate responsibility programs. Also, make sure to build in time for feedback. Given the SRI community's level of knowledge, their feedback can be extremely valuable for your program. By having an interactive discussion, you will build relationships with these important stakeholders.

The SRI community has formed an association known as SIRAN (the Social Investment Research Analysts Network). By connecting with SIRAN, your company can efficiently cover many of the SRI analysts with one meeting or conference call. Conducting group meetings with a variety of SRI representatives can be efficient, but it may not be sufficient for your company's needs. One-on-one meetings with SRI fund managers who are tracking your company are still the best way to have an in-depth dialogue. These one-on-one meetings will not only help solidify the relationship, but they also tend to allow for a more open exchange and a deeper dive into issues. In addition, the SRI funds and analyst groups all look at CR data in different ways. For example, some funds may focus on water usage statistics while others look primarily at labor issues in the supply chain. By meeting with a select group of fund managers and analysts you can build an understanding of their top issues, and hopefully address them at the meeting and follow-up.

Conclusion

At its essence, corporate responsibility is about integrating social and environmental concerns into corporate decision-making. To do this effectively, you need to understand the views of influential stakeholders external to your company who are well versed in these topics. While there are many ways to engage stakeholders and gain their perspectives, all leading corporate responsibility programs include some form of stakeholder engagement. Regardless of the format that works for your company, this is an essential element to building a credible program. Use the tips and experiences in this chapter to create a stakeholder engagement process that is customized to your company's culture and circumstances.

11

Employee engagement

This chapter outlines the methods for engaging your company's employees in corporate responsibility to achieve tangible business benefits.

Passion is the genesis of genius (Tony Robbins).

While it may not be intuitive or obvious, one of the biggest benefits that the corporate responsibility department can deliver to a company is employee engagement. Most people entering this field are focused on the benefits that their work can produce for people and the planet, but not necessarily motivating and inspiring their fellow employees. As you will read in this chapter, there is a very strong case for employee engagement as being one of the primary value propositions for the corporate treehugger.

Let's start by defining employee engagement. A human resources (HR) expert at AMD defined employee engagement as the "motivation to invest discretionary effort into work." According to Scarlett Surveys: "Employee engagement is a measurable degree of an employee's positive or negative emotional attachment to their job, colleagues and organization which profoundly influences their willingness to perform at work."[78]

You can boil all of this down into a simple statement: *engagement is seeing your job as your cause.* If your job is also your cause, you are naturally motivated to work hard. This is a view that most people who work in corporate responsibility can relate to because we are motivated by our altruism – our need to help people and the planet – and we have the good fortune to work for our cause.

The business value of employee engagement

There are reams of data that correlate employee engagement with business value:

- Companies with high engagement have three times the operating margin (profit) than companies with low engagement. In addition, 88% of fully engaged employees believe they can positively impact the quality of their organization's products and services, while only 38% of disengaged employees feel the same way[79]

- Organizations with more engaged employees have 2.6 times the earnings per share growth rate than organizations in the same industry whose employees are less engaged. Disengaged employees account for a loss of more than $300 billion in the productivity of the U.S. workforce[80]

There are many more studies that support the connection between employee engagement and business results. Because employee engagement is directly correlated to business success, it is the *raison d'être* for many corporate HR departments. These departments track their engagement scores on a routine basis by using employee surveys and will often develop new programs to address any deficits in the data.

How corporate responsibility drives employee engagement

What does CR have to do with employee engagement? Going back to my shortened definition (seeing your job as a cause), and knowing that increasing numbers of people (especially younger workers) care deeply about sustainability, there is a natural linkage between sustainability and engagement. There are several studies that link sustainability to engagement:

- The PricewaterhouseCoopers 2009 study *Managing Tomorrow's People. Millennials at Work: Perspectives from a New Generation* states that 88% of "millennials"[81] seek employers with values that match their own and 86% would consider leaving an employer whose values no longer reflect theirs

- The 2010 Hewitt Associates study *Engaging Employees through CSR* showed a strong correlation between the level of engagement and the percentage of employees who believe that their organization is socially and environmentally responsible[82]

- A 2008 study by Stanford Graduate School for Business surveyed 759 graduating MBAs at 11 top business schools; it showed that these future business leaders rank corporate social responsibility high on their list of values, and they are willing to sacrifice a significant part of their salaries to find an employer whose thinking is in sync with their own.[83] On average, these MBAs would give up a whopping 14.4% of their salary for a company that they perceived as a responsible corporate citizen

- In April 2011, the Society for Human Resource Management study, *Advancing Sustainability: HR's Role*, compared several measures of engagement in companies with superior sustainability programs against those with poor sustainability programs. The study reported the following benefits for companies with more engaged employees:
  - Employee morale: 55% improvement

  - Business processes: 43% more efficient
  - Public image: 43% stronger
  - Employee loyalty: 38% increased[84]

- In June 2011, *Fortune* magazine published a story titled "How Going Green Can Be a Boon to Corporate Recruiters," which summarized a number of studies: Mercer research found that workers under 25 years old listed a company's good reputation as fourth on the list of most important draw for a job, just under pay.[85] Buck Consultants reported that 49% of companies surveyed promoted their green agenda to attract environmentally conscious employees. Wayne Balta, IBM's Vice President for Environmental Affairs was quoted in the article: "They've figured out that companies that are progressive and innovating in this area are themselves innovative"

You don't have to be a genius to connect the dots here: engagement is about seeing work as a cause; sustainability is a cause for many people; therefore, great sustainability programs are correlated with superior employee engagement.

How to engage employees through corporate responsibility

With this model in mind, there is a clear business imperative to make sustainability a part of the work experience. The question is: how? Not everyone can work in the CR department (nor would they necessarily want to), but the data shows that current and potential employees are attracted to employers with outstanding sustainability and responsibility profiles. There are a number of ways that companies are engaging employees by integrating sustainability into their day jobs:

Green teams

Green teams are a growing trend with many companies (see Chapter 5 for a discussion on green teams). These are groups of employee volunteers that work together to improve the environment at work and in the local community. Green teams primarily focus on greening the work environment with activities that improve recycling, conserve energy, reduce cafeteria waste, and reduce commuting. Increasingly, these groups are taking on bigger challenges and responsibilities to the point where green teams are becoming a recognized organizational structure within some companies. Here are a few examples from company green teams that might spur some ideas for how you could work with your own company's green team:

Outreach to customers

The eBay green team started in 2007 as a grassroots effort to green the workplace. They went from eliminating Styrofoam cups to prompting eBay to build a large solar array. Then they had an epiphany: *Involve eBay customers*. This is an excerpt from the eBay green team webpage:

> As the world's largest marketplace for used, refurbished and vintage goods, eBay's biggest eco-friendly opportunity lies with its buyers and sellers – the millions of people that make up the eBay community worldwide. So in March 2009, we invited the eBay community of buyers and sellers to join in our green shopping efforts. After just six weeks, the Green Team was 100,000 members strong, and we're still growing. (We're pushing more than 225,000 now!) We want to harness the power of our community to help the world make smarter, greener shopping decisions. Join us in making a positive impact on the world (see: www.ebaygreenteam.com).

The exponential multiplier by engaging customers not only adds to the environmental impact of the eBay green team, but also enhances its brand reputation.

Dumpster days

To increase awareness about recycling, the employees at North Carolina-based Burt's Bees flipped over their dumpsters in the parking lot and separated out all of the recyclable items. According to 3blmedia.com, the purpose of the event was to educate employees about waste reduction. Trash destined for the landfill was divided into items that should have been recycled but were not, and garbage. With about five tons of stockpiled trash dumped onto the parking lot, employees donned HAZMAT suits and dove in to find out what they could dig up. They saved approximately 2.8 tons of trash from landfills. And, according to GreenBiz.com, the lessons learned from the dumpster-diving experiment resulted in a 50% reduction in waste, saving the company around $25,000 annually.

Personal sustainability plans

Walmart is the world's largest company with over 2 million employees (aka associates). This is how Walmart describes its employees' personal sustainability plans (from Treehugger.com):

> As sustainability has become a standard part of the next generation Walmart business model, our associates have begun making sustainability part of their daily lives. And they're very passionate about it. That's why we've introduced a new global program, My Sustainability Plan (MSP), to help engage and support our associates who want to choose an activity that will allow them to become more sustainable.
>
> Associates can choose as many goals as they like and track their progress online. In the US, they can visit our internal social networking site, mywalmart.com, to select and track their goals. This allows associates to connect with colleagues to provide encouragement and inspiration, or even spark some spirited competition among friends, stores, or even entire divisions. In addition, this format allows everyone to understand how their individual actions – when combined with those of their 2 million colleagues – add up to make a world of difference.

Biofeedback for buildings

At AMD, the new Lone Star campus in Austin, Texas is LEED Gold certified. While this certification means that the design features are among the greenest in the country, it does not guarantee that the operations of the building are also green. Enter the AMD green team. Like other green teams, these dedicated volunteers are working on campus-greening issues ranging from composting cafeteria waste to commuting alternatives. With the support of the facilities group, the green team (and others) recently got a new and powerful tool to manage onsite energy use: SmarteBuilding.com. This technology allows for real-time feedback of the energy use in buildings down to a small cluster of offices. You can watch the total energy use for your area and break it down by plugs, lights, and building temperature. Armed with this technology, the AMD green team is planning an "energy night out" to find out what energy-consuming equipment is left on when people leave the office. They are also working on ideas for setting up energy conservation competitions among various teams.

Environmental excellence awards

As discussed in Chapter 5, recognizing employees for their environmental protection efforts is a successful strategy for driving environmental improvements. This kind of recognition can also be an effective means for improving employee engagement.

When I first introduced this award at Intel, many of the nominations were for personal actions that had an environmental benefit, such as carpooling or recycling. As employees started to see the value that the company ascribed to these accomplishments, we started to get nominations for larger changes like improvements to business or manufacturing processes. We realized that this award had motivated a hidden army of eco-minded employees who had incorporated environmental improvements into their day jobs. While we did not measure their levels of employee engagement, it was fairly obvious that everyone involved in the process was excited about lightening the company's environmental footprint and took great pride in the recognition. To further bolster the employee engagement value of this process, I turned the management of the award over to the Intel green team (with funding and oversight from my group) and it became one of their top priorities.

Skill-based volunteering

This is an idea that has become a hot topic in the world of employee volunteerism. The concept is to match an employee's skill set and/or development needs to a volunteer opportunity. There are a number of groups that have this concept at the heart of their business model (e.g., yourcause.com and angelpoints.com). The HandsOn Network describes the concept this way: "Connecting the volunteer with the right skills to the right project at the right time will allow getting a greater impact and building stronger relationships between volunteers and the nonprofit sector."[86]

An essential element of this kind of volunteering is the matching system. Your company will need a way to identify the core skills of volunteers and match them with the skill needs of qualified nonprofit partners. Skills-based volunteerism is the perfect marriage between your HR department and the community affairs department. HR wants to see employees developing their skills and becoming engaged through volunteerism. The community affairs team wants to increase employee volunteering. By matching employees' skills to nonprofit needs, this method of organizing your volunteer program can benefit both.

Micro-volunteering

The number one reason people do not volunteer for a cause while at work is they do not have enough time. If you think about the typical volunteer experience, it does not fit nicely into a busy workday. Most people have multiple meetings and commitments on their calendar every day and, while they may want to get involved, they have not been able to free up a large enough chunk of time to make it happen. Enter the concept of micro-volunteering. It turns out that most people do have enough time to volunteer, but it is just cut up into small chunks.

A company called Sparked (sparked.com) has figured out a way to take advantage of these smaller chunks of time with an online application that matches employees' skills with nonprofit needs in an online environment. For example, if a nonprofit needs assistance with its marketing plan, it can post this need on the sparked.com system, which will match the need with participating company employees that have marketing skills. Instead of updating their Facebook page between meetings, these employees can log on sparked.com and help out a nonprofit with their marketing plans, brand image, copywriting, operations, Web design, and a host of other jobs. Because this can all be done from their desk, it is far more efficient and attractive to busy people. The system also allows all of the engagements to be tracked and monitored.

Micro-volunteering may get more employees involved in a cause, but it needs to be measured differently than traditional volunteering. The primary volunteerism metric is the number of hours invested in volunteer activities. Representatives from Sparked told me that the average duration of an engagement on their site is seven minutes. So, it is not likely that your volunteer hours will rise significantly by using this method. But, by counting the number of employees involved as a primary metric, you might find a significant increase in volunteerism, and thus engagement, by using this tool. While it will not replace the need for "hands-on" volunteers, this approach has the potential to rewrite the rules for skill-based employee volunteerism. The quote from National Public Radio on the Sparked website says it well: "crowdsourcing for the common good."

Ultimately, the ability of your CR program to engage employees, ranging from the executive office to the rank and file, is a primary determinant of success and a significant element of your value proposition as a corporate treehugger. A theme that runs through this book is partnering for success. Successful employee engagement is dependent on solid partnerships with the departments that manage your company's workforce. Depending on your company's structure, you should reach out to the HR team, the community affairs team, or both. Seek first to understand their programs and plans before suggesting any enhancements around sustainability/corporate responsibility. Again, it is best if you are able to get these ideas embedded into the functions that have line management responsibility for employee engagement (typically HR or community affairs) and then serve as a catalyst and advisor.

When you do suggest new ways to engage employees through your sustainability programs, be prepared to put in the time to get a new program off the ground. In most cases, even if your partner organizations are supportive, they will not have the time or resources to implement new ideas immediately. In my experience, if you can invest enough time, energy, and resources to get a program to launch with some interest and excitement, there will be a line of people wanting to take control (and credit) for the concept. Like a start-up company that gets bought out by a larger company, spinning off successful programs should be viewed as success. It means that the concept worked and now has an institutional home that will keep it alive long after you have moved on to other issues.

Working on employee programs is the definition of a win–win endeavor: by effectively engaging the employees of your company in sustainability programs, you can unlock huge benefits for the environment while simultaneously returning significant business value for your company.[87]

12

Diversity, governance, and ethics

This chapter delivers the essential concepts needed to under-stand and communicate the diversity, governance, and ethics aspects of your corporate responsibility program.

Action indeed is the sole medium of expression for ethics (Jane Addams).

Diversity: the art of thinking independently together (Malcolm Forbes).

So far this book has covered the "bread and butter" issues at the center of the corporate responsibility profession. But there are equally important issues that are included in the definition of a responsible company. If you study the Global Reporting Initiative (GRI) guidelines you will see a number of questions aimed at the diversity of the company's workforce, the governance structure (as well as compensation), and corporate ethics policies.

Like the previous chapters, I won't attempt to dive into each of these topics in great detail but instead provide an overview and practical tips that you can apply in real-world settings. Remember, the corporate responsibility manager has to work with a wide variety of functions that cut across multiple business units within the company. For most of these functions, you will have limited expertise and no direct authority over how the function is managed.

Diversity

Diversity in the workplace is a sensitive and complex topic and there is an entire library of books dedicated to the subject. Diversity is sensitive because it can appear to employees that the company is trying to practice social engineering on the workforce by favoring some groups over others to make up for past discriminatory practices. After all, the messages that most companies send to their employees is that they will be judged by the merit of their work not attributes they cannot control such as race, gender, or sexual preference.

So, why do so many companies have a diversity program? There are several answers to this question. The first is business value. In the library of diversity books you will find study after study showing that diverse groups of people produce better business results. Once this data started to enter the corporate consciousness, some companies started to count the numbers of their employees in each demographic category to see if they had a diverse workforce. Depending on the business, the numbers likely did not match the proportion of these groups in the general population – meaning that the company had been hiring a disproportionate number of a certain group compared with the demographics of the available hiring pool. Often, the lack of diversity is even greater at higher levels of management within the company (including the board of directors).

If the data are skewed toward one or more demographic groups, most companies will examine their hiring and promotion practices to ensure that there aren't any overt biases being applied. Even if no bias is discovered, many companies will seek to take steps to adjust the imbalance through affirmative action policies that give preference to underrepresented demographic categories such as females or people of color.

Because it is easy to count, tracking the proportion of each demographic in your workforce is the lowest common denominator for diversity management programs. For example, in AMD's latest corporate responsibility report the company disclosed the proportion of females in senior management (11%) and the percentage of females in the overall workforce broken out by region (ranging from 20 to 43% – fairly typical for a technology firm).

Effective diversity management is far more complex than counting percentages. There are many different categories of employee attributes that, when mixed together, can form a diverse and productive work culture. Diversity managers break down the mix this way:

- **Core (personality).** Companies are interested in the fundamental characteristics of employee behavior in the workplace. These core characteristics are often measured by personality inventories such as the Myers-Briggs Type Indicator® (MBTI). Understanding personality traits is useful because the data can indicate who is compatible with whom in workplace team situations. For example, when I had my team take the MBTI at Intel, the lesson was that the people on the team processed information very differently.[88] While we didn't take this exercise to the next level by rearranging work teams based on personality type, this very simple reminder helped us ensure that all voices are heard and included in team processes

- **Primary (age, ethnicity, gender, race, sexual orientation, or physical ability).** These are the traditional demographic definitions of diversity. These categories are important because each of these groups is a legally defined protected class – meaning that if discrimination against employees in these classes is proven, the company can be liable for damages and penalties. While many companies track data on how many people are in each class, most do not report it publicly. While the data is useful as a gross indicator of diversity, it does not tell you enough about what is really going on in the workplace

- **Secondary (religion, language, work style, education, work experience, relationship status, recreational habits, family/ parental status).** With the exception of religion, relationship, and family status, most of these attributes are not within legally protected classes and, thus, are almost never tracked in any formal sense. Nevertheless, these factors can be critical from a diversity and inclusion perspective. For example, there are strong cultural sensitivities regarding religion in some countries where your company may conduct business. Being aware of these sensitivities and managing them well is an important aspect of your company's diversity program

While there may appear to be an element of political correctness in diversity management, the ultimate goal of the diversity and inclusion program is to get the most value out of your company's workforce. Again, the data suggesting that there is a business value in diversity and inclusion is strong. For example, the top 50 companies in Diversity Inc.'s Top 50 have a 25% greater shareholder return than the S&P 500 average.[89] Catalyst (catalyst.org) reported that companies with the most women in top management have a 35% higher return on equity and a 34% higher total return to shareholders than the companies with the fewest women in top management.[90]

AMD's diversity director, Deb Nations, told me that "diversity is not about counting heads, it's about making heads count." What she meant is that diversity practice has matured from quota-based management (counting heads within each group) to finding ways to include diverse perspectives from across the spectrum of employee characteristics. It is about building employee relationships to improve group performance and productivity, and fostering the flexibility and the agility needed to adapt to changing markets and consumer preferences. Consumer product companies in particular have figured out that if their workforce – and especially their management – does not match the overall demographics of their customer base, then they are likely to misunderstand their market and lose business.

Truly effective diversity programs look at their workplace demographics *and* how the people in their workforce are treated. In other words, *diversity is the mix, while inclusion is making the mix work.* While a diverse employee population might look good in your corporate responsibility report, the real business value is unlocked when the broad range of backgrounds, experience, and skills are included and their creativity is leveraged into business decisions.[91]

Governance

Governance refers to how your company is managed, including its structure, policies, executives, and directors. The items of importance from a corporate responsibility standpoint are:

Independence

Investors are particularly interested in whether the board is independent from the executive management of your company, meaning that they exercise meaningful oversight over the finances, strategies, and operations of a company, rather than simply rubber-stamping management's plans. Several GRI questions are focused on board independence: for example, "State the number of (board) members that are independent or non-executive" and "Indicate whether the Chair (of the Board) is also an executive officer." The New York Stock Exchange has established specific standards for the independence of board members.[92]

Environmental and social governance (ESG)

The ESG concept is based on the belief that if the charter and structure of your company includes environmental and social issues, the company is more likely to achieve progress on these issues.

This concept is illustrated by a 2010 shareholder resolution filed by Harrington Investments to amend Intel Corporation's bylaws to create a board committee on sustainability. Intel initially opposed the resolution but eventually agreed to change the charter of one of its board committees (Governance and Nominating) to include oversight of sustainability. The theory is that the additional attention and oversight from a board committee will improve the company's level of sustainability performance. Notably, the GRI guidelines ask companies to disclose which board committees have "direct responsibility for economic, social and environmental performance."

The 'B' corporation

On January 1, 2012 a new law in the State of California came into effect that created a new corporate governance structure allowing corporate directors to consider social and environmental considerations on equal footing with financial returns. Before this law came into effect, shareholders could sue corporate boards if the company took on costly environmental or social initiatives instead of maximizing profits.[93]

The companies that incorporate under these laws are often called "B" corporations, which is short for "benefit" corporations. Similar laws have now passed in six other U.S. states and companies are beginning to sign up.[94]

In California, the designation under the new law is called the "Flexible Purpose Corporation." Flexible purpose corporations write a social or environmental mission(s) into their articles of incorporation. For example, one company may be interested in preserving water supplies, while another may set its mission to eliminate labor abuses in its supply chain. The law requires the company's directors to consider these social or environmental missions their decision-making, even when it could mean lower returns for investors.

One of the first companies to sign up was Patagonia – maker of outdoor clothing and gear. Patagonia has a long tradition of promoting environmental and social causes and created a big splash in late 2011 with its "common threads" advertising campaign. In a full-page *New York Times* ad, the company asked consumers not to buy their products if they could reuse or recycle older gear.

My friend, and longtime proponent of the benefit corporation structure, Suz Mac Cormac (Partner with the San Francisco law firm Morrison & Foerster), is working with several companies beyond Patagonia that are considering utilizing the flexible purpose corporation option. Whether or not the "B" corporation takes off with major brands is still unknown, but it presents an attractive alternative for corporate governance that allows social and environmental considerations equal standing in corporate governance.

Policies

To demonstrate strong governance, most companies will list the relevant corporate policies in their corporate responsibility reports. Examples of these policies include: the principals of corporate governance, standards of business conduct, code of business ethics, stock ownership guidelines, and the supplier code of conduct. Each of these policies sets out the company's expectations with regard to its employees' and suppliers' conduct involving ethical, social, and environmental issues. In many instances, legal requirements such as the Sarbanes–Oxley Act, the Foreign Corrupt Practices Act, and a long list of labor laws dictate the content of these policies. It is important to understand all of your company's policies that pertain to social, environmental, and ethical concerns.

Procedures

Policies are only effective if they are actually implemented and their compliance is monitored. Corporate responsibility reports will usually include a list of business practices and corporate structures aimed at implementing governance policies. Examples of these procedures and structures include: the audit committee of the board; the internal audit department; the internal controls, compliance, and investigations organization; the risk management and business continuity office; employee alert lines and the processes to respond, investigate, and resolve complaints as well as whistle-blower anonymity and anti-retaliation procedures. One of the more common ethics procedures is mandatory training. Most companies will require ethics training and track compliance with the requirement.

Probably because of increased scrutiny, many companies now have fairly robust structures with sizable teams and established business practices to ensure conformance to ethics laws and company ethics policies. Ensure that you have a firm grasp of these procedures so that you are able to communicate them as a facet of your company's overall corporate responsibility program.

Stakeholder engagement

Chapter 10 delves into the practice of stakeholder engagement, but this topic is also included in most governance questionnaires. For example, in the governance and engagement section, the GRI guidelines ask companies to describe "Key topics and concerns that have been raised through stakeholder engagement, and how the organization has responded to those topics and concerns, including through its reporting."[95]

The concept behind this area of governance is the inclusion of stakeholder perspectives into business decisions. Leading companies in CR have specific procedures for including stakeholder views within a defined scope of ESG issues.

Governance is one of the topics that the corporate treehugger has to know about but typically has very little influence over. The decisions regarding corporate governance usually take place within the highest echelons of the corporation. The good news is that you will get to interface with these senior people as you seek to understand and report on the governance policies and structures of your company.

Another bright spot for the corporate treehugger in the governance area is the increased focus on sustainability from the board of directors and senior management. As the above example from Intel demonstrates, many external stakeholders are demanding that company leadership have more awareness and ownership of corporate responsibility issues. In practice, this means that you may be asked to assemble information into presentations summarizing your programs, progress, and plans for senior management and/or the board. Some companies have made this into a formal function within the executive ranks but, as of today, these companies are a minority. Most boards and executive teams are focused on the more traditional measures of business success: profit and loss, business strategy, investors, customers, products and services, and research and development.

If you are called upon to interface with executive management and/or the board on corporate responsibility, work on making the information relevant to their paradigm. First, think through your main messages and boil them down to their essence. Start your presentation or written summary with the key messages and then fill in the details if appropriate. Remember, people at this level will have little time or attention span for a lot of background or detail. Another tip is to tie the corporate responsibility information to relevant business issues. Porter and Kramer's 2006 article "Strategy and Society"[96] is an informative guide to why and how corporate responsibility is more effective when it is tied to the core competencies of the business. Senior leadership will likely glaze over, or worse, become antagonistic, if you present your program as "the right thing to do" or simple charity. Every issue you present should have a clear, compelling, and quantitative (if possible) benefit for the business.

Ethics

The term "corporate ethics" will often elicit a few laughs as an oxymoron. People tend to think that companies are driven only by the profit motive and will do whatever it takes to make money without regard to values or morals.

Like most stereotypes, this view exists because it is based on a kernel of truth. If you look under the hood, most companies have probably cut some corners or skated close to the edge of the law at some point to conduct their business. As mentioned earlier, there are a slew of legal and regulatory requirements that have been issued in response to notorious ethical lapses. The Sarbanes–Oxley Act – issued in the wake of the Enron collapse – protects corporate whistle-blowers and requires companies to establish employee hotlines to report fraud. The Foreign Corrupt Practices Act regulates and requires reporting of "facilitation payments," which are part of doing business in many countries around the world. The recent Dodd–Frank Wall Street Reform and Consumer Protection Act seeks to protect investors against the excesses of the financial community that were so evident in the financial crash of 2008–2009. And these are only a few of the regulations and requirements focused on corporate ethics.

So, with all of these laws and regulations in place, how does a corporate treehugger deal with ethics issues? First, take inventory of the items you will need to report in your corporate responsibility report. For example, the GRI guidelines ask for the "percentage of employees trained in the organization's anti-corruption policies and procedures."[97]

After you assemble all of the GRI questions that concern corporate governance and ethics, set up meetings with your company's corporate ethics officer and corporate secretary to discuss how they may want to respond to these. You may find that the ethics officer is interested in telling a more holistic story behind the ethics policies and practices of your company. Many companies have made substantial investments in this area and will be justifiably proud of their programs. For example, AMD has adopted a leadership role with regard to the issue of fair and open competition as an outcome of the company's struggles with Intel, and this program has been a feature of the annual CR report for several years.

There are several ranking and rating schemes that focus on ethical performance such as EthicalQuote by Covalence (ethicalquote.com), Ethisphere's World's Most Ethical Companies, and the Ethibel Sustainability Index. The issues covered by these ethics ratings include:

- Product liability
- Antitrust
- Bribery
- Harassment and discrimination

- Government contracting

- False claims; marketing and branding

- Environmental, health, and safety violations

In the past, most companies were not overly candid about ethical lapses and violations. Increasingly, however, companies are self-disclosing ethical lapses. There are real incentives for self-disclosure as most of the applicable laws contain steep penalties for unreported "known violations" – where the company knew there was an issue but did not disclose it to the appropriate authorities.

Of course there are plenty of gray areas in business ethics. For example, a sales team may be under pressure to sell enough products so that the company can achieve its revenue target before the close of the quarter. This team might use leverage with a customer to "pull-in" some orders before the end of the quarter. Is this an ethical violation? The answer to many questions like this is, "it depends." This is why many companies now have a chief ethics officer to make these types of determination on a real-time basis. While many companies report on their ethics policies, investigations, enforcement structures, and training in their corporate responsibility report, fewer are also reporting on the number, severity, and resolution of violations.

The corporate treehugger's job is to understand your company's ethics policies, procedures, and performance, and to represent this information to external stakeholders. It is a great idea to start the ethics section of your report with a statement from a senior corporate officer with responsibility for ethics (your company's general counsel is ideal). This statement will reinforce the importance of ethical behavior as an expectation emanating from the top of the company. It is critical that these leaders also "walk the talk." Even small lapses can create the perception of inattention to ethical issues, which may lead people to believe that they have tacit license to cut corners. Part of your job in corporate responsibility is to communicate a strong culture of ethics and compliance internally and externally to your company.

Conclusion

A former boss once said to me "one 'aw crap' trumps 1,000 'attaboys'."
Ethical lapses can crush a company's reputation and overwhelm all of
your hard work on corporate responsibility issues. The press loves a
scandal and will relentlessly hound a company when there have been
real or perceived ethical lapses (examples include: the HP board and
CEO scandals, Toyota's safety issues, the BP oil spill and safety issues,
Intel's antitrust violations, and Goldman Sachs's behavior in the mortgage meltdown). While the corporate treehugger will not typically
have a direct role in these issues, it is crucial that you understand your
company's ethics reputation and adapt your communications and programs accordingly.

13

Recognition, awards, and rankings

This chapter delves into the complex, but important, world of corporate responsibility rankings, lists, ratings, and awards.

Don't worry when you are not recognized, but strive to be worthy of recognition (Abraham Lincoln).

Allow me to start this chapter with a bit of a rant: The field of corporate responsibility has become a ratings game. On the one hand, this can be viewed as a positive force as companies compete for top billing in the various public ratings. On the other hand, the ratings game has become something of a beauty contest that, at its worst, threatens the credibility of the corporate responsibility movement and, at its best, takes a sizable amount of corporate treehugger time. For a deep dive on this topic, I recommend the "Rate the Raters" study by Sustain-Ability mentioned in Chapter 10. It is a comprehensive and timely exposé on the issues associated with the proliferation of CR ratings.

A recently launched initiative called the Global Initiative for Sustainability Ratings (GISR) is aimed at bringing consistency to how companies are ranked on responsibility issues. Its goal is to create a framework for the convergence and harmonization of the vast number of sustainability ratings that have cropped up over the years. Its task will not be an easy one because sustainability ratings have become fairly entrenched in the corporate responsibility world.

Types of ranking and rating

Because there are many CR ratings and rankings, and because most companies take great pride in being recognized for their good deeds, working with the ratings organizations has become a major part of the corporate treehugger job. While I won't dig into any of the specific ratings and rankings, this chapter categorizes them and lays out some tips on maximizing your effectiveness to receive public recognition for your CR program.

Socially responsible investor ratings

Socially responsible investor (SRI) ratings refer to the analyses of company CR performance used to select and screen stocks. I list this category first because it may be the most impactful for your program. As discussed in Chapter 10, the latest estimate of assets invested using social management is an eye-popping $3.07 trillion, or nearly one out of every eight dollars under professional management, and growing ten times faster than the overall managed asset pool. As this area of investment has grown, the imperative to attract social investment has become a high priority for many corporate responsibility programs. As described in Chapter 10 the SRI community can be divided into SRI analysts and SRI fund managers. While their traditional customers have been investment fund managers, many of the SRI analyst firms have found additional clients by supplying their data and analyses to corporate responsibility rankings and ratings. For example, the ESG analyst firm Trucost supplies the data for *Newsweek* magazine's "Green Rankings" list. IW Financial is the data engine behind *Corporate*

Responsibility (CR) Magazine's "100 Best Corporate Citizens" list. The most well-known SRI ranking is the Dow Jones Sustainability Index (DJSI). Produced by the Sustainable Asset Management Group (SAM), the DJSI ranked 523 companies in four geographic regions of the world in 2011. The DJSI 2011 press release stated the purpose for the ranking this way:

> The DJSI follow a best-in-class approach, including companies across all industries that outperform their peers in numerous sustainability metrics. Each year, SAM invites the world's 2,500 largest companies, measured by free-float market capitalization, from the 57 sectors to report on their sustainability performance. The result of the Corporate Sustainability Assessment provides an in-depth analysis of economic, environmental and social criteria, such as corporate governance, water-related risks and stakeholder relations, with a special focus on industry-specific risks and opportunities.

There are many similar rankings and the number is growing every year.[98] The "Rate the Raters" study mentioned above inventoried over 100 responsibility rankings and ratings and is the most comprehensive study of this field to date.

Corporate responsibility lists and awards

There has been a recent explosion in the number of awards recognizing corporate responsibility. Other than a healthy balance sheet, companies like nothing better than to be recognized for excellence in corporate responsibility. Not only does this kind of recognition make everyone in the company feel good, but it also carries tangible business value.

I gave a talk at the 2011 Responsible Business Summit hosted by Ethical Corporation, and asked the audience, "Do companies compete on corporate responsibility?" I got an equivocal response when about half of the hands went up, so I said, "Okay, let's look at some ratings." On the next slide I showed the list of the top five companies on *Newsweek*'s "2010 Green Rankings" (Dell, HP, IBM, Johnson & Johnson, and

Intel). Then I added a graphic from the Reputation Institute, which showed that over 40% of a company's reputation stems from its corporate responsibility programs. The point was made with the final table, which showed the 2011 valuation of company brands from Millward Brown, which was led by Apple with a brand valued at $153 billion.

Here is the equation: corporate responsibility is about half of company reputation; reputation is a big part of your company's brand value; your company's brand value is the largest intangible asset on the company balance sheet. Do companies compete over their CR reputations? You bet they do. And one sure way to burnish your reputation is by being named in a lot of corporate responsibility awards and rankings.

A first step to approaching CR awards and rankings is to understand how the ratings systems evaluate your performance. For example, AMD appeared on *Corporate Responsibility Magazine*'s "100 Best Corporate Citizens" list in 2010 and 2011. This is one of the more transparent rankings, which gave me some insight into some of the issues inherent in their systems.

When AMD was initially listed, the company ranked first in overall governance and ethics. The next year, AMD was still on the overall Top 100 list, but was ranked 270th in governance and ethics. Nothing had changed in AMD's programs, so I was perplexed about how this score could have dropped so much. When we talked to the *CR Magazine* folks, we learned that they had changed their grading system to "Olympic scoring." This meant that 269 companies tied for first and all received a grade of one. For unknown reasons, AMD was dropped into the second group and the score dropped from first to 270th. This example gives you a sense of the somewhat random ways that your company will be judged by the various rankings and ratings schemes.[99]

By definition, sustainability is a broad concept that incorporates many disparate issues ranging from profitability to diversity and from environmental protection to ethics and fair labor practices. The notion that you can aggregate these items into a single score or into a meaningful ordinal ranking is nonsensical. The key for the CR manager is to understand the various rankings schemes and put your best foot forward. For example, some rankings may put more weight on environmental performance than labor issues, while others focus on human rights or ethics. Understanding these weightings and matching them with your company's strongest stories will give you a guide for where to invest your time.

Great place to work lists

At the other end of the rankings and ratings spectrum is the plethora of "great place to work" lists. The best known is the Fortune® "100 Best Companies to Work For" list which is compiled from data collected and ranked by the Great Place to Work® Institute.

There are also a myriad of specialty lists. For example, *Working Mother* magazine publishes several lists that rank companies such as:

- Working Mother 100 Best Companies List

- Best Companies for Multicultural Women

- Best Companies for Women's Advancement

- Best Companies for Hourly Workers

- Best Green Companies for America's Children

Another important list is the "Corporate Equality Index" from the Human Rights Campaign, which lists the top companies that support equality for LGBT (lesbian, gay, bisexual, and transgender) employees.

You could fill up a whole chapter with all of the "best places to work" lists, but the critical thing to understand is how best to approach these lists. As with the responsibility rankings discussed above, focus on a few of the lists with the highest return on your investment. The "best places" lists are usually based on a survey of your company's employees. While this method is arguably more realistic than relying on company-provided information, there are some issues you need to consider before signing up. The surveys can be very long and require a large number of employees to take time to fill them out. Collectively, this can be a considerable investment of resources into the survey which, when added to all the other surveys your company issues, may be too burdensome.

Also, because the responses are beyond your control and the scoring methods are often not transparent, you have little control or insight into the final results. Regardless of your ranking, these surveys can be useful to evaluate the strengths and shortcomings in your programs. Most companies prefer to use a professional survey service for assessing employee satisfaction and engagement rather than relying on a "great place to work" survey. While you get some feedback from a "great place to work" survey, you may not be able to extract enough detail from the data to really understand what is going on in the workforce.

Managing SRI funds and analysts

The best way to approach SRI analysts and funds is to pick out a few companies and get to know them. You should work with your investor relations department to create a list of the SRI funds that already own shares in your company as a starting point. Then, take an inventory of the SRI funds that don't currently invest in your company and prioritize a few of them as potential investors. Next, add the SRI analysts that are influential with the funds you have targeted. From the combined list, create a schedule by which you and your IR representative will contact each entity on the list either by visiting in person or by phone. In these meetings, you will present an overview of your corporate responsibility progress and plans, but remember to leave plenty of time for feedback and interaction (see the section in Chapter 10 on the investor road show). Your goal in these meetings is to learn as much as you can about the issues of importance to the fund managers and analysts so that you can address them in your program.

Managing corporate responsibility rankings

How can you effectively work within the world of rankings and ratings? Because many of the rankings depend on publicly available information, recognition depends on being able to tell your company's story in a clear and compelling way to a focused audience.

Start by creating a list of the awards and rankings that your company currently receives. Identify those that you want to continue participating in as well as new awards that you believe are within your reach and would be impactful for your company. Next, dig into the "data engines" behind the awards. Since many corporate responsibility awards rely on the same analyses as the SRI funds, you may already have relationships with the SRI analysts that evaluate your performance for the awards you selected. Make a list of all the analysts that research your company and make it a priority to speak to each of them to walk through your story, answer questions, and provide the background behind the data. You will also need to track down the award or ranking process and discover as much as possible about the weighting and scoring system. Get to know the people at the conferring organization to understand their process and let them know you are interested in their award.

The bottom line is that the criteria and process for being selected for awards can be subjective, non-transparent, and somewhat frustrating. There will be awards your company will receive that will be a happy surprise, and awards panels that reject your company for unsatisfactory reasons. Stay focused on your priority list of awards and avoid the temptation to fill out every award application and survey. Do a re-evaluation of your target awards at least annually to revise your list. It can be hard to stay focused when you get e-mails about a competitor that is listed for an award that you did not receive.

When you review your target list, do a sober evaluation of the awards and rankings that your program has a reasonable chance to attain. For example, it may become clear that you will never break in to the Dow Jones Sustainability World Index, but you have a great shot at the 100 Best Corporate Citizens list. If it is clear that you are wasting time with an award or list, consider reallocating the resources to an award that you might have a better chance of winning. While there is always a "big fish, little pond" trade-off, it can be a good strategy to look for awards where you can stand out as a leader rather than being an runner-up to better funded programs.

OneReport® is a service that aggregates the survey questions from many of the major ESG research organizations into a single online questionnaire that can be parsed into topic areas and sent to the various data providers within your company for completion. While OneReport® is an excellent tool, the sheer number of ESG surveys means that the aggregated list contains more than 1,500 questions. Completing and reviewing all of the questions requires a tremendous amount of work and can stress your network of data providers. If you use this service, remember that all of the effort you invest in filling out the survey is still not sufficient to assure that your program will stand out. In today's world, there is increasing competition for a place on CR ratings or rankings lists and the best way to get your story out is to tell it yourself.

There is no substitute for getting in a room with the analysts face to face (or at least on the phone) to walk through your company's CR story. The personal storytelling approach will allow you to build relationships with your counterparts in research and ranking firms, and permit you to talk through the story behind the data. For example, when AMD transferred major manufacturing assets to a joint venture, it impacted the overall environmental footprint of the company. Through a series of discussions, AMD helped the analysts understand the new business model and explained how the company had adjusted its environmental and CR programs.

Rankings and ratings matter, but they have to be kept in context. Remember that, while external recognition is great, it is not the sole reason that you are working in corporate responsibility. The important elements to remember about rankings and ratings are:

- Prioritize and focus on those the rankings with the highest applicability and return on investment for your company

- Tell stories rather than just dumping data

- Build relationships with your analysts; and, above all

- Remember that ratings are often subjective and may not be representative of your company's actual performance

As a final word, try not to get too wrapped up in the ratings and rankings world. While filling out surveys and managing everyone's expectations will take up large amounts of your time, the majority of your bandwidth should be focused on developing programs that continuously improve your company's genuine performance measures.

14

Match your passion to your profession

This chapter forecasts the trajectory of the CR profession and provides practical tips for getting a job and lessons to take on your journey.

> It is up to us to live up to the legacy that was left for us, and to leave a legacy that is worthy of our children and of future generations (Christine Gregoire).

Today, corporate responsibility is a growing profession with a career ladder extending into the executive suite of the world's largest companies. This is a relatively recent change and the knowledge, skills, and abilities for this emerging field are still being defined. The CR role cuts across almost all business functions and is housed in a wide variety of corporate departments. Similarly, there is enormous variability in the job responsibilities, ranging from the art of communications, to the technically demanding field of environmental management.

The newness and variability in this profession means that there are many different backgrounds that can qualify for and thrive in these roles. Notwithstanding this variability, throughout this book I have described the common skills and attributes essential for success in corporate responsibility. This book has also delved into the substantive aspects of the role that you must know to acquire a job and succeed in the field of corporate responsibility.

Corporate responsibility is, in fact, a real job with real responsibilities, unique pressures, and serious demands. While rewarding on many levels, the job of a corporate treehugger can be frustrating, because you will always be something of a "stranger in a strange land" – meaning that many of your colleagues will not understand what you do or how your work adds value to the company. When you work in this field, you will have to perfect your 30-second "elevator speech" about what is corporate responsibility, and how your role enhances the company brand and its attractiveness to employees and investors. At times you may feel like a metronome – vacillating between the euphoria stemming from your laudable accomplishments and the dejection from the feeling that your role in the corporate power structure sits somewhere between superfluous overhead and oblivion. There have been many moments in my corporate treehugger career when I was on top of the world and many other times when the late comedian Rodney Danger-field's catchphrase summed it up well: "I get no respect."

Even if you can stomach the bipolar nature of this role, you may be challenged by its breadth (which is very wide) and depth (which can be very shallow). This job requires that you are knowledgeable and – to some extent – responsible for a wide range of corporate behaviors and programs for which you have almost no control. And when you do achieve great accomplishments in this field, you should almost always deflect the credit to others within the company. If this were not bad enough, you should also consider that, until quite recently, corporate responsibility was widely considered a career dead end. Today, while this career is not considered the fast track to the executive suite, there is a positive trajectory. Many companies now have a senior vice president or vice president of corporate responsibility, and that list is growing with the increased awareness of CR issues. In addition, companies are realizing that many of the skills needed to perform well in corporate responsibility are portable to other career paths.

There are an increasing number of people who are interested in working in corporate responsibility departments or contributing to this growing field from their chosen profession. The academic and literary fields have responded to this demand by pumping out an amazing number of courses and books aimed at preparing people for these careers. The problem is that almost all of these instructional aides are written and delivered by people with little experience actually doing these jobs. Carefully researched business case studies and rhetorical arguments about why and how corporations could and should contribute to society are wonderful, but ultimately not all that helpful to getting and keeping a job in corporate responsibility.

As the mega multinational corporations gain importance in our collective zeitgeist, they are increasingly subject to our approval and acceptance of their operations. I believe that the market will ultimately punish the companies that seek only to take money out of our wallets without equal or greater contributions to our well-being. As high-profile examples of corporate screw-ups and public apologies mount – from Enron to BP – you don't need a marketing PhD to figure out that a good reputation is vital to a company's long-term success. And increasingly, the public is tuned into corporate responsibility as a major factor of a company's reputation.

As this reality seeped into the corporate consciousness, it has stoked a CR competition among big companies. Traditional economic theory would predict that companies would always seek to maximize return on investment; indeed, there is a strong argument that this is the fiduciary responsibility of corporate officers and directors. As globalization has taken hold and traditional barriers have shrunk, this theory would predict that companies would move their operations to take advantage of cheap labor and lax regulation. While there is little doubt that these forces are at play, the "race to the bottom" is being countered by competition over CR reputation. The increased awareness of corporate responsibility (as manifested by the proliferation of ratings, rankings, and socially screened investments), as well as the values and expectations of younger employees, are palpable signs of the competition in responsibility or "race to the top" that is a fact of life for today's big companies.

My own career reflects this change. I started off working on regulations dealing with local environmental impacts. Later, I was enforcing social and environmental standards with Apple's suppliers in countries around the world. Now, at AMD, I am working to end the human rights abuses in the dangerous minefields of the Democratic Republic of Congo where the profits from mining "conflict minerals" are funding some of the worst human rights abuses of our generation. In just a couple of decades, the scope of corporate responsibility has grown from local to global, from single impacts to multiple impacts, and from a single company to a whole supply chain. This field is still evolving, and with this evolution new career opportunities are being created.

Getting a job in corporate responsibility

While the boundaries of corporate responsibility are still expanding, so is the profession. A few years ago, I would have said that this role is too much of a niche to support real job growth. Now, I see a field that is growing in opportunities and stature:

- The world's largest company, Walmart, employs a senior vice president of sustainability, and VP-level corporate responsibility positions exist or are being created at many of the Fortune 500 companies

- The essential skills for this job have been mapped by several organizations (The Boston College Center for Corporate Citizenship website has one of the most comprehensive catalogues of relevant papers on the essential skills for these jobs.)

- The Corporate Responsibility Officer Association is developing a "code of practice" for the profession

- The International Organization for Standardization published ISO 26000, outlining standardized management systems for the practice

- The Reputation Institute estimates that more than 40% of a company's reputation stems from its corporate responsibility practices and communications

The evidence is mounting that corporate responsibility is a necessary function with staying power. The rise of the corporate responsibility profession has been compared to the quality movement in the 1980s. During this period United States manufacturers were concerned about losing their edge to Japanese companies that were delivering high-quality products at lower costs. This concern spawned an entire movement that gave rise to quality departments in most major companies, and entire new disciplines, like Six Sigma, that are now integral to many companies' operations.

While the quality movement is mature, the CR field is still evolving and more jobs are being created. In my experience, many of these jobs are initially created from inside the company. Once the executive decision is reached to initiate the CR function, companies will typically find seasoned managers from within their ranks to fill these spots. Because these roles have to work across so many functions within the company, it makes sense to select a leader who has familiarity with the business. Also, some companies use the cross-functional nature of the role as a good training ground for executives they are grooming for other positions.

As this field matures, the tendency to appoint corporate responsibility managers from within will change. Increasingly, companies are seeking experienced professionals who come fully loaded with all the capabilities and connections needed to vault their programs into a leadership position. Some of the major headhunting firms (executive recruiters) have formed specialties around these positions and there are a growing number of firms and websites devoted to placement in this profession (e.g., see Martha Montag Brown, Ellen Weinreb, BSR. org, sustainablebusiness.com/jobs, netimpact.org/do-good-work/job-board, Greenjobs.com, brightgreentalent.com, and others). The book, *Profession and Purpose* by Katie Kross (Greenleaf Publishing, 2009), is also an excellent reference for corporate responsibility job hunters.

Most companies launch their CR departments with a senior leader who will then assemble a staff by hiring entry- and mid-level positions. Currently, many companies have small CR departments – typically a senior person and one or two staff members. CR departments are beginning to expand as sub-specialties are being defined. For example, communications and marketing is a major sub-element needed to manage corporate responsibility reports, websites, updates, blogs, rankings, ratings, and events. There are an increasing number of entry-level and mid-level jobs within corporations, consultancies, or PR firms to help manage this workload.

As discussed in Chapter 6, the field of supplier responsibility is an increasing job market for experts in this area. Outsourcing and greater public awareness of social compliance will continue to feed this market for years to come. These are excellent jobs for people interested in getting deep into the tactical operations of companies and supply chains. They are also great jobs for driving tangible improvements for people and the planet.

Getting started

You are coming to the end of this book, you have researched the field, and now you want to go out and get a corporate responsibility job. What do you do? Where do you start?

Start with a bit of self-reflection and analysis. If you were completely honest with yourself, would you be best suited for a more technical role (e.g., supply chain auditor), a less technical role (e.g., communications) or a managerial position (e.g., corporate responsibility director or vice president)? Use your self-analysis to filter (or at least prioritize) the jobs in your search.

Next, think about the companies in your search. In Chapter 1 we categorized companies as "2x4" (those who have been whacked and now see the value in CR) and "epiphany" (where CR is an integral part of the company's mission and business model). Almost any job in an "epiphany" company will give you exposure to corporate responsibility – which can open up many job possibilities – but these companies are typically overwhelmed with résumés and the competition is fierce. Corporate responsibility jobs within "2x4" companies are more focused and thus less plentiful, but there are many more companies in this category.

Consider the maturity of the corporate responsibility program within your target companies. More mature programs are likely to have more jobs, but the jobs will also be more specialized and thus constrained to certain aspects of the program. Jobs in less mature programs will be more entrepreneurial but also more ambiguous and chaotic. In these programs you may find yourself designing the strategy and developing the programs. If you go to work in one of these programs, you should be comfortable dealing with ambiguity and being self-directed. Another factor to watch out for in your job search is under-funded programs. As mentioned above, many corporate responsibility departments are small – often just a leader and one or two people. These roles will require you to cover a wide swath of the issues and responsibilities outlined in this book. To the extent feasible, and in the interest of your own self-preservation, you should seek clarity on the scope, responsibilities, and objectives in these roles.

Interviewing tips

When you narrow it down to the companies and jobs you will go after, there are a few tips that will help you in the selection process:

Passion

I am continually surprised when I interview candidates and they don't take the opportunity to tell me about why they would be thrilled to work in the role. The interview is your chance to express your passion for the work and you should absolutely volunteer this information. It conveys a sense of engagement to the hiring manager. As discussed in Chapter 11, people who see their work as a cause are far more likely to go the extra mile. I like to hear a personal story about why this is an important field for you and why you chose to invest time in this career path.

Prove your passion

Show concrete evidence of your passion. It's not nearly enough to go into an interview and say you care about the environment if the most you can claim is that you recycle your soda cans. Volunteering – in a way that shows real commitment and produces real accomplishments – goes a long way to making this case. Hiring managers tend to be skeptical about candidates who say they are committed to an issue but cannot present evidence that they support the cause.

Experience

If you are entering the workforce, a CR-related internship will give you a definite boost in the job hunt. Hiring managers like to see you've taken this step to gain experience in the field. If you don't have a directly related experience, draft a résumé and cover letter that shows how your experiences can specifically add value to the CR role in question. For example, if the role is about communications, demonstrate your experiences and capabilities in communications that would be directly applicable to this role. I recommend developing two or more versions of your résumé that highlight certain experiences along the lines of the jobs you are seeking. You should also draft a specific cover letter for each job that customizes your personal story and experience to demonstrate how you could add value in the position. Use your cover letter to tell your story in your own personal way. A résumé can only convey so much, so your cover letter must weave the relevant threads through your experiences and your personal goals and passions that led you to apply for this position at this time.

Research

In the Internet age, it astonishes me how many candidates show up with limited knowledge of the company, the role, and the hiring manager.

In a couple of minutes, you can find out reams of data online. In a few more minutes, you could probably link up with a few insiders and get the scoop about the job and company. If you are considering working at a company for years, doesn't it make sense to spend a few minutes on the Web to discover what you can? Someone once asked me if I was put off when candidates reveal that they have researched my background during an interview. For me, it is a net positive: it means that the candidate has done his or her homework and come prepared. While most hiring managers will be flattered if you talk about accomplishments, don't overdo it – remember, the hiring manager is assessing you.

Practice your pitch

There are a few standard techniques that interviewers will use to assess candidates and, since you have only one hour or less to make an impression, it makes sense to practice your story and your responses in advance. For example, Intel uses a standard technique called "behavioral interviewing." This method assesses your capabilities by asking you to discuss real-world experiences about specific situations. A typical question on how you manage ambiguity could be: "Tell me about a time when you were given a vague assignment. What was the assignment, what did you do to clarify the goals and how did it turn out?" There are references you can find on the Web about this technique that can help you prepare. Select a few of the skill areas that you think are relevant for the role and develop your examples in advance. Also develop your opening remarks that tell your story: who you are, your passion for the company and the role, and how your experience and personal capabilities will add value. I recommend practicing in a mock interview with a friend – preferably one who has experience interviewing candidates.

Advice for career changers

A 2008 Ethical Performance salary survey of CR professionals in Europe showed that more than half had changed careers to enter the field. If you're a career-changer, highlight how your skills and previous experience can transfer into the new, CR-focused position. For example, while at Apple I hired a person (Kirsty Stevenson) with no CR experience (but loads of passion for the role) from the supply chain organization specifically for her skills, contacts, and experience in managing suppliers. Show that you bring skills to the table that can advance the needs of the program while telling the story of why CR is your calling. For career-changers in particular, you can come to the interview with a compelling story of why your values have led you to the decision to change the course of your career to corporate responsibility.

Lessons to take on your journey

Throughout this book I have covered the skills, competencies, and attributes that will help you with your career in corporate responsibility. As you head out on this journey, here are a few more lessons I have learned along the way that I hope are helpful to you:

Understand the business

Chapter 1 opens with the statement "the business of business is business." While perhaps trite, the message is that companies are not charities. For-profit enterprises survive or fail based on the value they deliver to the market. As a CR worker, you need to understand your company's value proposition on a deep level and articulate how and why your CR programs add to that value.

Don't get too far in front of the cavalry

Because the rest of the business is focused and busy delivering a product or service, it is easy to fall into the delusion that you are the final authority on decisions about CR. You need to understand where your program fits in the overall power structure of the organization and adjust accordingly. For example, there are a myriad of CR principles and pledges (e.g., the UN Global Compact) that your company will be asked to endorse. You may be able to make a case for or against adopting one of these pledges, but the ultimate decision must reflect the policy of the corporation. It is a huge mistake to make commitments on behalf of your company without lining up the needed support (cavalry) and going through the appropriate decision processes.

All ideas are good ideas until you have to pay for them

Because the field of corporate responsibility is new, in many companies you will likely get involved in strategy and program development. As your program grows and matures, you will need to choose the areas that you will invest in carefully. With a scope as broad as corporate responsibility, there are nearly unlimited possibilities. An important skill for your success is to analytically differentiate the merely good ideas from the ideas that have the highest leverage for your program. Chapter 3 discusses the criteria and process for finding the right programs for investment.

Don't take yourself too seriously

Benjamin and Rosamund Stone Zander, in their book *The Art of Possibility*, coined the term "rule number 6," which is totally meaningless unless you know the definition that the Zanders ascribe. Benjamin Zander used "rule number 6" as a code phrase to remind his students not to take themselves too seriously. Great advice in my opinion. In my career, I have seen many people allow their personalities to become enmeshed with their positions. After a while, they begin to believe in their own hype and become arrogant or inflexible. Always remember that, regardless of your stellar résumé and accomplishments, you have been hired for your capabilities not your reputation. Avoid "silo" behavior where you are more concerned about your own performance or your group's performance than the overall benefits to the company. Always seek out ways to help others, and your effort will earn interest and come back to you with dividends.

Words matter

This is a phrase I learned from the Chief Counsel for the Senate Environment and Public Works Committee, Mike Evans, when we were working through the legislative language. It is a fairly obvious observation but it stuck with me because, too often, we assume that others will know what we mean. While this assumption is especially dangerous when you are writing legislation, it can be just as relevant in the practice of corporate responsibility. You will be asked to communicate on a wide variety of corporate programs and policies, many of them complex and sensitive. Often, you will have to communicate these complex, sensitive topics to a broad audience of stakeholders who may not understand all of the details, but are still outraged and emotional about the topic. As outlined in Chapters 8 and 9, the ability to communicate clearly and concisely is an essential skill for the corporate treehugger. If you are not a good writer or speaker, make it a priority to strengthen these skills. These are high leverage skills for corporate responsibility and a wide variety of other vocations.

Results matter more

The field of sustainability has become a "gabfest" of conferences, webcasts, blogs, websites, and forums extolling the shift of business toward solving social and environmental problems. If we could tap into the resources and energy expended on *talking* about sustainability and use a fraction of it to *achieve* sustainability, we could make real progress. While you will participate in many of these events and communications about sustainability and corporate responsibility, always ask yourself, "how is this activity helping to make things better?" If the answer is elusive or unclear, think about using your time on another project that has a clearer direct benefit to people and the planet.

Culture eats strategy for breakfast

This is another catchphrase that stuck with me because it is true. Every organization I have worked in has a distinct and different culture. I am not sure how corporate culture gets started and maintained, but I do know that every grouping of human beings who are together for any length of time will develop a unique culture – a way that they relate to one another. Just as knowing your company's business model is essential, so it is essential to understand your company's culture. In Chapter 2, we delved into "reading the system" and how different companies behave very differently. This behavior – or culture – will set the tone and context of your corporate responsibility program. For example, some companies will endorse every corporate responsibility standard, join every sustainability group, and speak or sponsor every major CR conference. Other companies would prefer to quietly work behind the scenes. Setting a strategy that is divorced from the reality of the culture of your organization is a fool's errand.

The imposter syndrome

This is a known psychological phenomenon in which people are unable to internalize their knowledge and accomplishments. In other words, sometimes you feel like you are an imposter in your position and people may discover the "real you" at any moment. I believe that people in corporate responsibility jobs are especially prone to this syndrome because they have stepped away from whatever discipline or specialty they had established as the center of their career and are constantly dealing with issues in which they have little experience. This is the opposite of "don't take yourself too seriously"; the advice here is to remember that you probably know more about corporate responsibility than most of the people you will encounter. While you always want to be curious, try to balance this with a quiet confidence about your knowledge, capabilities, and contributions.

Lead from wherever you stand

As stated in the Introduction, you may find yourself in a job where you have limited authority to use the lessons from this book because your role is constrained to a narrow scope. Never let your position supersede your passion or overshadow your abilities. Sure, you might not be the CEO yet, but you can look for ways to suggest changes or enhancements. Meet with senior leaders to understand their issues and problems. Be a curious, lifelong learner, an innovator, and a problem solver. Going the extra mile to help solve a problem is not only a rewarding experience, but it will also demonstrate your commitment to others in the workplace. All great leaders who started from lower-level positions discovered ways to leverage their capabilities to add value to their organization. Use the ideas in this book to enhance your program regardless of your position in the organization.

Final thoughts

There are just a few more reflections and observations to close this book and send you on your way into the world of corporate responsibility. As mentioned several times in these pages, many of the best practices for a career in CR are applicable to other careers. It is reinforcing to know that the investment you put into growing your skills for a career in CR can pay off in other career paths. Many of the skills covered in this book, such as reading the system, emotional intelligence, leading through influence, communications skills, being results-oriented, self-aware, self-analytical, and self-motivated are all ingredients for success in a variety of careers. Also, assuming the role of a servant leader – leading by giving priority to the needs of your organization and the colleagues you serve – is a recipe for success in any role. Remember that your personal credibility, your knowledge, and the trust you evoke in others are tremendous assets that cannot be traded away.

While this book is focused on a career in corporate responsibility, you can contribute to social and environmental good from any job within a company. Most CR departments are clearinghouses for information. They collect and communicate the data and accomplishments from other functions in other departments. If you are working in one of the traditional functions within a company, explore the opportunities to create a wonderful story for your colleagues in the CR department. For example, if you work in procurement you could leverage your core competency in supplier management to implement a green purchasing program. If you are in the real estate group, you could drive LEED building specifications or purchase renewable energy credits. If you are in human resources, you could work on initiating or enhancing your company's diversity and inclusion programs. The list is nearly endless.

The noted humanitarian Dr. Paul Farmer[100] – a Harvard-educated physician who has devoted his career to providing health care to the poorest people in the world – responded to a question about what advice he would give to college students who want to change the world with this statement:

> Acknowledge the world the way it is. Acknowledge your privilege. Understand that there are things that can be done and that there is not one path but many if we want to take on these problems. [Ask yourself] what is it that you like to do . . . [and then apply these skills] to promote the movement for social justice globally.

These are wise words from a good man. The choice to work in corporate responsibility is one way, but certainly not the only way, to promote the movement for social justice and environmental protection. Corporate responsibility acknowledges the world the way it is, dominated by companies with revenues and global power beyond the wildest dreams of most nation-states. This book is dedicated to helping you, the corporate treehugger, develop the competencies to achieve social and environmental improvement through a career within a for-profit enterprise.

Endnotes

Introduction

1 The term corporate responsibility (CR) is used synonymously with the term corporate social responsibility (CSR).

2 A. Karnani, "The Case Against Corporate Social Responsibility," *Wall Street Journal*, 23 August 2010.

3 CorporateRegister.com, "Stats," www.corporateregister.com/stats, accessed January 21, 2012.

4 The word "footprint" is a popular term used to describe the total inventory of an environmental impact, through the lifecycle of a product or process. The term "carbon" is a euphemism for total greenhouse gas emissions. Thus, a "carbon footprint" is the total greenhouse gas impacts through the lifecycle of a product or process.

5 Sarbanes–Oxley Act. Pub. Law. 107-204.

6 Dodd–Frank Act. Pub Law. 101-203.

7 UN Commission on Environment and Development, *Our Common Future* (Oxford, UK: Oxford University Press, 1987).

8 A. Smith, *The Theory of Moral Sentiments* (Edinburgh: A. Kincaid and J. Bell; London: A. Millar, 1759).

9 C. Christensen, "How Will You Measure Your Life?" *Harvard Business Review* 88.7–8 (July 2010): 45-61.

Chapter 1

10 E. Weinreb, "CSR Jobs Report 2010: 2004–2009 Longitudinal Study Conclusions," April 7, 2010, weinrebgroup.com/insights/csr-jobs-report-2010, accessed July 16, 2011.

11 S.R. Covey, *The 7 Habits of Highly Effective People: Restoring the Character Ethic* (New York: Free Press, 2004).

12 Landor Associates; Penn, Schoen, Berland; Cohn & Wolfe; and Esty Environmental Partners, "The 2011 Image Power Green Brands Survey," June 2011, www.cohnwolfe.com/en/ideas-insights/white-papers/green-brands-survey-2011, accessed January 10, 2012.

13 "The Luddites were a social movement of nineteenth-century English textile artisans who protested – often by destroying mechanized looms – against the changes produced by the Industrial Revolution, which they felt were leaving them without work and changing their way of life. The movement was named after General Ned Ludd or King Ludd, a mythical figure who, like Robin Hood, was reputed to live in Sherwood Forest" (en.wikipedia.org/wiki/Luddite, accessed January 26, 2012).

14 M. Gladwell, *The Tipping Point: How Little Things Can Make a Big Difference* (Boston, MA: Little Brown & Co., 2000).

15 E. Lipton, M. McIntire, and D. Van Natta Jr., "Top Corporations Aid U.S. Chamber of Commerce Campaign," *New York Times,* October 22, 2010.

16 The term comes from the Brundtland Commission report *Our Common Future* (1987), which defined sustainable development as "Development that meets the needs of the present without compromising the ability of future generations to meet their own needs." Five years later, the 1992 Rio Declaration on Environment and Development – or "Rio Declaration" – was issued, which consisted of 27 principles intended to "guide future *sustainable development* around the world."

17 The number of transistors that can be placed inexpensively on an integrated circuit doubles approximately every two years (Gordon Moore, "Cramming More Components onto Integrated Circuits," *Electronics Magazine*, April 1965).

18 Visionaries like Bill Sheppard pioneered "the right thing to do" culture at Intel. For an excellent case study of Intel's sometimes rocky road during this period, see "Growing Pains: Rio Rancho

Wooed Industry and Got It, Plus Financial Woes," *Wall Street Journal*, April 11, 1995: A1.

19 In practice, the term "sustainability" has become almost synonymous with environmental issues. Because the term "corporate social responsibility" includes the word "social," some people may equate this to labor and human rights issues. Many companies have adopted the term "corporate responsibility" to be inclusive of both environmental and social issues.

20 From www.globalreporting.org: The Global Reporting Initiative (GRI) is a network-based organization that pioneered the world's most widely used sustainability reporting framework. GRI is committed to the Framework's continuous improvement and application worldwide. GRI's core goals include the mainstreaming of disclosure on environmental, social and governance performance.

21 CorporateRegister.com, "Stats."

22 The International Organization for Standardization (ISO) has published a guideline on corporate social responsibility titled ISO 26000: 2010 "Guidance on Social Responsibility." In addition, The Boston College Center for Corporate Citizenship published an excellent resource titled "Profile of the Profession 2010: Corporate Citizenship Leaders for Today and Tomorrow" (June 25, 2010); see: www.bcccc.net/index.cfm?fuseaction=document.showDocu mentByID&DocumentID=1387 (accessed January 11, 2012). Lastly, the Corporate Responsibility Officer Association published a draft code of ethics for corporate responsibility managers: "CROA: Draft Ethics Code" (March 16, 2010); see: www.croassociation.org/node/838/16571 (accessed January 11, 2012).

23 A recent Stanford Business School Survey concluded that new MBAs would give up more than 14% of their starting salary to work within a sustainable business. D.B. Montgomery and C.A. Ramus, *Calibrating MBA Job Preferences* (Working Paper; Stanford, CA: Stanford Graduate School of Business, 2008).

Chapter 2

24 This lesson is reminiscent of the parable of "boiling the frog." If you put a frog in hot water, he jumps right out. But if you put him

in at a comfortable temperature and turn up the heat one degree at a time, he will stay there until he is cooked.

25 S. Mol, *Classical Fighting Arts of Japan: A Complete Guide to Koryū Jūjutsu* (Tokyo, Japan: Kodansha International, 2001): 24-54.

26 *Ibid.*

27 The designation "skunk works" is widely used in business, engineering, and technical fields to describe a group within an organization given a high degree of autonomy and unhampered by bureaucracy, tasked with working on advanced or secret projects.

28 M.E. Porter and M.R. Kramer, "Strategy and Society: The Link between Competitive Advantage and Corporate Social Responsibility," *Harvard Business Review* 84.12 (December 2006).

29 M.E. Porter and M.R. Kramer, "The Big Idea: Creating Shared Value," *Harvard Business Review* 89.2 (January 2011): 62-77.

30 C. Christensen, *The Innovator's Dilemma: The Revolutionary Book that Will Change the Way You Do Business* (New York: HarperCollins, 2000).

31 Much like the dashboard in your car tells you about the essential functions on a real-time basis, a management dashboard tracks key indicators that determine the health of a business function.

Chapter 3

32 From www.globalreporting.org. See note 20. The GRI has become the de facto international standard of the issues and key performance indicators that define corporate responsibility.

33 SAM (Sustainable Asset Management), "Dow Jones Sustainability Indexes," 2011, www.sustainability-index.com, accessed August 22, 2011.

34 Open-ended questions are an important business tool but can be hard to master. A good example of the difference between an open- and closed-ended question is asking someone: "is this issue a priority for your business?" vs. "tell me about the issues that are a priority for your business." The first question could elicit a yes or no response; the second question allows the person to explain their issues in their own words. You will learn more with open-

ended questions and the conversation will be far more engaging. Practice this skill with a friend and count how many times you end up asking a question that can be answered with a yes or no answer.

35 Business for Social Responsibility has done an outstanding job conducting these analyses for me at three different companies.

36 Ceres is an organization (www.ceres.org) that specializes in stakeholder engagement on corporate responsibility and can manage the details for you.

37 One of the more innovative ways to attain stakeholder input on a materiality analysis is the interactive matrix on SAP's website. This matrix allows anyone from outside the company to move issues between the four quadrants based on their own perspectives, then tabulates and updates the average of all input.

38 A popular exercise to underscore the advantage of team processes is to set out a problem and ask individuals to first try to solve the problem their own; then do the same exercise in groups. Invariably, group processes produce better results because of the diversity of experiences and perspectives.

39 There are many "icebreakers" for getting people used to participating at the start of meetings. Many are personal questions such as "tell us about one hidden talent." I tend to like icebreakers that are tied to the topic of the day, such as asking each participant: "what are your expectations for this meeting?"

40 There is something magical about writing someone's words on an old-fashioned flip chart. It allows people to "see" and internalize the thoughts and increases their ability to build on the concepts. Words that are not captured on flip charts often float away as soon as the next person starts to talk. Drawing pictures, tables and diagrams also helps participants visualize the progress of their discussion. Again, a good facilitator can add value to group meetings with techniques to elicit and capture the group's creativity.

41 The most concise and memorable vision statement I have heard came from a business case on the Xerox Corporation. The vision statement was "beat Canon."

42 Be selective in choosing KPIs. Not everything that can be measured should be. The act of measuring takes resources, which are multiplied when actions are taken to manage the indicator toward a goal. A good indicator *indicates* – i.e., it tells you something useful about the overall system being tracked without having to

measure the whole system. For example, personal space (in square meters) can be a good indicator of conditions in a supplier's dormitory without knowing much more about the dorm. Most indicators are lagging – telling you what has already happened. If possible, find some indicators that are leading or predictive. A good example of a leading indicator is repeat safety or environmental incidents – if you see the same issues again and again, there is likely a gap in your management system that could predict a larger failure in the future.

43 I *do* remember the day that Steve Jobs gave this speech. It was moving, but what got the crowd on their feet was when Jobs announced that all employees would receive a new iPhone (they were still pretty rare at that point).

44 O. Harari, *The Leadership Secrets of Colin Powell* (New York: McGraw-Hill, 2002).

45 United Nations Framework Convention on Climate Change, UN Doc FCCC/CP/1997/7/Add.1, December 10, 1997; 37 ILM 22 (1998).

46 At a scenario planning session I helped set up at Intel, we hired a graphic artist to sketch out the alternative scenarios in pictures and color them in based on how the group reacted. Although this kind of support is not essential for scenario planning, it was a fun and engaging process for all involved.

47 Here are a few basic hints for staying on top of the growing mountain of information on corporate responsibility topics: download "TweetDeck" or "HootSuite" to keep track of a variety of social media feeds and hashtags such as #csr and #sustainability on one screen. Establish a few "Google Alerts" for important issues – Google's search engine will gather and e-mail relevant Web hits for your search strings at regular intervals. Sign up for "report alert" to get real-time notification of when new CSR reports are issued.

Chapter 4

48   Coca-Cola Company, "Live Positively," 2011, www. thecoca-colacompany.com/citizenship/index.html, accessed August 22, 2011.

49 Starbucks, "Starbucks Global Responsibility Report: Goals & Progress 2010," 2011, www.starbucks.com/responsibility/learn-more/goals-and-progress, accessed January 22, 2012.

50 Kaplan and Norton's *The Balanced Scorecard* is the bible for many corporate managers for the development of management dashboards. It is worth a read, but it is my experience that, while you can draw useful lessons from this book, a literal application is not a perfect fit for most corporate responsibility programs.

51 When we were establishing this program at Intel, my wife used to mark the week before an operations review on her calendar because she knew that I would be largely unavailable.

52 I recall a meeting where Tim Cook asked me about an NGO report on Apple suppliers. I was surprised that he was even aware of the report. He not only knew about the report, he had read the report and had a question about a claim made on a page that was deep in the document.

53 Apple does not have a corporate responsibility function per se. The main functions are divided into the applicable business units. While this structure reduces bureaucracy, it also limits the coordination and communication between functions with overlapping interests.

Chapter 5

54 The term "environmental footprint" is used to describe the overall environmental impact of a product or process. The term is often applied to the lifecycle of a product and it can be sub-categorized by certain issues such as the water footprint or climate (or "carbon") footprint.

55 K. Kleiner, "The Corporate Race to Cut Carbon," *Nature Reports Climate Change* doi:10.1038/climate.2007.31, August 2007.

56 See Peter Sandman's risk communication framework at www.psandman.com (accessed January 10, 2012).

57 This does not mean that you should ignore actual risks that have yet to make it into the public consciousness. The point is that you need to ensure that all known or anticipated health and environmental risks are well managed *as well as* address the perception of risk.

Chapter 6

58 The smiling curve illustrates value added (return on investment) as a function of the product lifecycle. The curve shows that the highest returns are in the beginning and the end of the product lifecycle. There are slimmer profits in the middle of the lifecycle – assembly and manufacturing. This observation has become even more pronounced with the acceleration of globalization in recent years.

59 J. Dedrick, K.L. Kraemer, and T. Tsai, *ACER: An IT Company Learning to Use Information Technology to Compete* (Irvine, CA: Center for Research on Information Technology and Organization, University of California, Irvine, 1999).

60 For more information about Social Accountability International, visit www.sa-intl.org (accessed January 10, 2012).

61 Read about the ISM Principles of Sustainability and Social Responsibility at www.ism.ws/sr/content.cfm?itemnumber=18497 &navitemnumber=18499 (accessed January 10, 2012).

Chapter 8

62 Toastmasters International, www.toastmasters.org, accessed October 1, 2011.

63 *Gallup Management Journal*, "Gallup Study: Engaged Employees Inspire Company Innovation" (Washington, DC: Gallup Organization, 2006; gmj.gallup.com/content/24880/gallup-study-engaged-employees-inspire-company.aspx, accessed August 22, 2011).

64 Hewitt Associates, LLC, *2010 Best Employers in Canada Study* (Vancouver, BC: Hewitt Associates, 2010; www.hewitt.com/bestemployerscanada, accessed August 19, 2011).

65 Social Investment Forum Foundation, *2010 Report on Socially Responsible Investing Trends in the United States* (Washington, DC: US Social Investment Forum, 2010; ussif.org/resources/research/documents/2010TrendsES.pdf, accessed August 21, 2011).

66 BrandZ Rankings, "BrandZ Global Top 100," 2011, www.
    millwardbrown.com/BrandZ/default.aspx, accessed October 1,
    2011.

67 Karnani, "The Case Against Corporate Social Responsibility," *Wall
    Street Journal*, 23 August 2010.

Chapter 9

68 See the GRI reporting statistics at www.globalreporting.org
    (accessed January 10, 2012).

Chapter 10

69 The Future 500, "Sustainability Toolkit," 2011, www.future500.
    org/sustainability-toolkit, accessed February 1, 2012.

70 To learn more about the XL program and the policy please read:
    "The Alternative Compliance Model: A Bridge to the Future of
    Environmental Management" by Timothy Mohin in *Environmen-
    tal Law Reporter* 27.7 (July 1997).

71 See the matrix at: www.sapsustainabilityreport.com/be-heard
    (accessed January 10, 2012).

72 *Investor Relations Weekly* reported in May of 2011 that there are
    108 entities creating corporate sustainability ratings, but that
    fewer than 20 are used by the average company.

73 This report is in four parts and provided free of charge on the Sus-
    tainAbility website: www.sustainability.com (accessed January 17,
    2012).

74 Social Investment Forum Foundation, *Report on Socially Respon-
    sible Investing Trends in the United States*.

75 *Ibid.*

76 *Ibid.*

77 AMD has taken a leadership stance on this issue by co-chairing
    a multi-stakeholder policy group with an NGO called Project
    Enough (a project of the Center for American Progress).

42   Chapter 11

78 Scarlett Surveys International, "Blog: What Is Employee Engagement?" 2011, www.scarlettsurveys.com, accessed August 19, 2011.

79 Towers & Watson Employee Surveys, *Employee Surveys: Talent and Rewards* (New York: Towers & Watson, 2011; www.towerswatson.com/services/Employee-Surveys, accessed August 22, 2011).

80 *Gallup Management Journal*, "Gallup Study: Engaged Employees Inspire Company Innovation."

81 PricewaterhouseCoopers, *Managing Tomorrow's People: Millennials at Work: Perspectives from a New Generation* (London: PwC, 2008; www.pwc.com/gx/en/managing-tomorrows-people/future-of-work/pdf/mtp-millennials-at-work.pdf, accessed August 21, 2011).

82 Hewitt Associates, "2010 Best Employers in Canada Study."

83 Stanford GSB News, *Challenging Work and Corporate Responsibility Will Lure MBA Grads* (Stanford, CA: Stanford GSB, 2008; www.gsb.stanford.edu/news/research/montgomery_mba.html, accessed August 22, 2011).

84 Society for Human Resources Management, *Advancing Sustainability: HR's Role* (Alexandria, VA: SHRM, 2011; www.shrm.org/Research/SurveyFindings/Articles/Pages/AdvancingSustainability HR%E2%80%99sRole.aspx, accessed August 21, 2011).

85 S. Du Bois, "How Going Green Can Be a Boon to Corporate Recruiters," *Fortune*, June 2011, tech.fortune.cnn.com/2011/06/02/how-going-green-can-be-a-boon-to-corporate-recruiters, accessed August 21, 2011.

86 See www.handsonnetwork.org/nationalprograms/skillsbasedvolunteering, accessed January 21, 2012.

87 Read Elaine Cohen's book, *CSR for HR: A Necessary Partnership for Advancing Responsible Business Practices* (Sheffield, UK: Greenleaf Publishing, 2010), for an in-depth look at engaging employees in corporate responsibility.

Chapter 12

88 In my experience, companies tend to hire and promote similar personality types. My belief is that company culture tends to select for certain traits which, over time, leads to homogeneity of the personalities in any particular group (birds of a feather flock together).

89 See DiversityInc Top 50 at www.diversityinc.com and the Corporate Leadership Council at www.clc.executiveboard.com (accessed January 10, 2012).

90 See www.catalyst.org (accessed January 10, 2012).

91 In addition to the employee diversity function (which is typically housed in the human resources department), many companies have an entirely separate program on supplier diversity. These programs were created because U.S. government procurement laws require a certain percentage of purchasing from certified, minority-owned, women-owned, and small businesses. Companies that have significant sales to the U.S. government will usually have someone in charge of monitoring supplier diversity within the purchasing department.

92 NYSE-Sponsored Commission on Corporate Governance, "NYSE Euronext Corporate Governance Guidelines," 2010, www.nyse.com/pdfs/CorpGovGuidelines_4-5-07.pdf, accessed August 21, 2011.

93 J. Tozzi, "Patagonia Road Tests New Sustainability Legal Status," Bloomberg.com, January 2012, www.bloomberg.com/news/2012-01-04/patagonia-road-tests-new-sustainability-legal-status.html, accessed January 23, 2012.

94 There is confusion between the voluntary "B" corporation pledge that some companies have taken (see www.bcorporation.net) and these new laws. The laws are far more impactful in that they set out the rules and liabilities for the corporation's governance, thereby creating the legal protections needed to factor environmental and social considerations into corporate decision-making.

95 Global Reporting Initiative, "Sustainability Reporting Guidelines," Version 3.1, 2000–2011, Section 4, https://www.globalreporting.org/resourcelibrary/G3.1-Guidelines-Incl-Technical-Protocol.pdf, accessed January 22, 2012.

96 M.E. Porter and M.R. Kramer, "Strategy and Society: The Link between Competitive Advantage and Corporate Social Responsibility," *Harvard Business Review* 84.12 (December 2006).

97 Global Reporting Initiative, *Indicator Protocols Set: Society* (SO, Version 3.1) (Amsterdam, NE: GRI, 2006; https://www.globalreporting.org/resourcelibrary/G3.1-Society-Indicator-Protocol.pdf, accessed January 22, 2012.

Chapter 13

98 *Investor Relations Weekly* reported in May of 2011 that there are 108 entities creating corporate sustainability ratings, but that less than 20 are used by the average company.
99 Regardless of their odd scoring system, the 100 Best Corporate Citizens list uses one of the most transparent ranking methods on the market. Other rankings do not disclose their methods.

Chapter 14

100 Read more about Dr. Farmer in the book *Mountains Beyond Mountains. Healing the World: The Quest of Dr. Paul Farmer* by Tracy Kidder (Random House, 2003).

Index

Milton Keynes UK
Ingram Content Group UK Ltd.
UKHW030902141024
449569UK00026B/1324